Religion and the Environment

Ralph Tanner and Colin Mitchell

Foreword by Vernon Reynolds
Professor
Department of Biological Anthropology
University of Oxford

palgrave

First published 2002 by
PALGRAVE
Houndmills, Basingstoke, Hampshire RG21 6XS and
175 Fifth Avenue, New York, N. Y. 10010
Companies and representatives throughout the world

PALGRAVE is the new global academic imprint of St. Martin's Press LLC Scholarly and Reference Division and Palgrave Publishers Ltd (formerly Macmillan Press Ltd).

ISBN 0–333–91974–2

This book is printed on paper suitable for recycling and made from fully managed and sustained forest sources.

A catalogue record for this book is available from the British Library.

Library of Congress Cataloging-in-Publication Data

Tanner, R. E. S. (Ralph E. S.), 1921–
Religion and the environment/by Ralph Tanner and Colin Mitchell.
p. cm.
Includes bibliographical references and index.
ISBN 0–333–91974–2 (cloth)
1. Human ecology—Religious aspects. I. Mitchell, Colin, 1927– II. Title.

GF80 .T36 2001
215′.7—dc21

2001040651

10 9 8 7 6 5 4 3 2 1
11 10 09 08 07 06 05 04 03 02

Printed and bound in Great Britain by
Antony Rowe Ltd, Chippenham, Wiltshire

Contents

List of Figures viii

Foreword ix

Preface x

1 Introduction: Religion and the Environment 1
- 1.1 Introduction and definitions 1
- 1.2 The difficulty of recording religious activities 2
- 1.3 The scope of the religion/environment interaction 3
- 1.4 The scale of the religion/environment interaction 4
- 1.5 The effects of religious doctrines 7
- 1.6 The intensity of the human/environment interaction 8
- 1.7 The degree of religious centralization 10
- 1.8 The degree of religious exclusivity 12
- 1.9 Direct versus indirect effects 13
- 1.10 The effects of religious beliefs on long-term environmental planning 15

2 Types of Religions 17
- 2.1 The four main types of religion 17
- 2.2 The religious practices of ancient societies 18
- 2.3 Small-scale tribal and traditional religions 22
- 2.4 Chinese and Japanese religions 24
- 2.5 Hinduism, Buddhism, Jainism, Sikhism, and Zoroastrianism 28
- 2.6 Judaism, Christianity and Islam 39

3 The Limitations on Religion/Environment Interactions 45
- 3.1 The incompleteness of religious observance 45
- 3.2 Inter-religious tensions 46
- 3.3 The physical constraints on religious practices 50

4 Perceptions of the Environment 53
- 4.1 The universality of environmental perceptions 53
- 4.2 Religious awareness of the environment 54
- 4.3 The influence of antecedent beliefs 55
- 4.4 The afterlife and attitudes to the environment 58
- 4.5 The role of wealth 63

4.6 Linguistic and cultural limitations on environmental understanding 65
4.7 Religious art and the environment 67

5 The Effect of the Environment on Religions 72
5.1 General environmental effects 72
5.2 The effect of homogeneous environments 75
5.3 The effect on Hinduism 77
5.4 The effect on Buddhism 79
5.5 The effect on Christianity 79
5.6 The effect on Islam 81
5.7 Social and physical constraints on religious practice 82
5.8 The effects of modern industrial and post-industrial environments on religion 84
5.9 The effects of population increases on religion 84

6 The Impact of Religious Behaviour on the Environment 86
6.1 Enthusiasm and drive 86
6.2 The use of time 86
6.3 The use of space 88
6.4 The use of materials 91
6.5 The use of transport 96
6.6 Domestic religious practices and the environment 99
6.7 The use of foods 104
6.8 The environmental effects of communal religious buildings 107
6.9 Outdoor shrines and the environment 110
6.10 The environmental effects of religious cities 114
6.11 The routeways used by religious practitioners 118
6.12 The natural environment and sacred spaces 122

7 Socio-economic Activities and the Environment 126
7.1 The impact of changing socio-economic conditions on religion 126
7.2 Religion and specific occupations 127
7.3 States and societies with one dominating religion 131
7.4 States with segmented religious systems 136
7.5 The environment in religiously divided societies 138
7.6 The effect of religion on agriculture 140
7.7 Settlement patterns 146

8 Religious Influence and Population Pressures 150
8.1 Factors connecting religion and population levels 150

8.2 Psychological factors in religion and population pressures 151
8.3 Inter- and intra-society religious violence 153

9 Religious Acts with Consequences on Communal Lifestyle 160
9.1 Reducing morbidity and mortality 160
9.2 Increasing morbidity and mortality 163
9.3 Care of the short-term disabled 167
9.4 Care of the illegitimate, orphans, and the long-term disabled 170

10 Religious Influences on the Life Cycle and the Maintenance of Environmental Balance 172
10.1 The chances of conception 172
10.2 The chances of survival into adulthood 176
10.3 Marital arrangements 180
10.4 The disposal of the dead 182
10.5 The influences of the dead on the living 186
10.6 The prohibition of contraception 188
10.7 Violent religious practices reducing pressure on the environment 190

11 The Impact of Religion on the Treatment of Animals 193
11.1 The direct and indirect effects of religion on animal life 193
11.2 The effects of religion on particular animals 196
11.3 The use of animals in sacrifices 200

12 Religion and Environmental Management 204
12.1 Secular authorities and religious activities 204
12.2 Current religious activities and environmental management 205
12.3 Future concern of religions for the environment 207

13 Religion and the Environment: an Overview 209

Bibliography 212

Index 232

List of Figures

1.1	Buddhist shrines in Chiang-Mai, Thailand	6
1.2	Half-burnt bodies on banks of the Ganges at Banaras	13
2.1	Aztec Pyramid of the Sun, Teotihuacan	19
2.2	Western India. Buddhist cave carved out of rock	20
2.3	Stone images of Buddha in Myanmar	21
2.4	Tanganyika, Usukuma. Ancestor shrines in compound	23
2.5	Japan. Deer in temple grounds in Nara	26
2.6	Hindu India. Cow in streets of Jaipur	30
2.7	Myanmar. Buddhist monks on their morning begging round	34
2.8	Hindu India. Monkeys in temple	35
2.9	Bangkok, Thailand. Dogs in temple grounds	35
3.1	Turkish Cyprus. Destruction of Christian cemetery	47
3.2	Christian use of materials, Rouen cathedral	51
4.1	Hindus bathing in the Ganges	57
4.2	London. Highgate cemetery as nature reserve	60
4.3	Croatia. Flowers on Roman Catholic graves on All Souls' Day	61
4.4	Moslem tombs near Cairo	62
4.5	Ruins of temple at Karnak, Egypt	65
4.6	Contemporary Hindu poster with environmental themes	70
6.1	Memorials in cemetery near Tokyo, Japan	95
6.2	Cremation container for body, Bali, Indonesia	96
6.3	Buddhist shrine for ancestors in Chiang-Mai, Thailand	101
6.4	Candles burning at Buddhist shrine, Beijing	102
6.5	Buddhist ordination of trees in Chiang-Mai, Thailand	103
6.6	Pagoda at Chiang-Mai, Thailand	108
6.7	Roman Catholic blessing of mountain peak	111
6.8	Myanmar. Buddhist pagoda on boulder	111
6.9	Lourdes grotto with crutches of those miraculously cured	112
6.10	Wailing Wall, Jerusalem. Jews praying	114
6.11	Buddhist mobile shrines used in holy day parades in Sri Lanka	115
10.1	Parsi Towers of Silence in which the dead are abandoned	183

Foreword

All religions influence the environment to a greater or lesser extent. All religions promote attitudes to the physical environment, the birds, animals and landscape that surround us. In this book the authors focus on these effects. For example, if we take traditional Judaeo-Christian ideas as represented in the books of the Old Testament the overriding concept is that God created the earth for man to use in ways that were useful to him. He was given 'dominion' over the earth and the beasts. Some species were 'taboo', and we find in many religions that certain species of birds and animals are taboo and not to be eaten or killed. In such ways religions act to conserve the environment. But in general the mood of Christianity is to make man the master of the environment and in modern society where pollution and rape of the environment are common, it has been claimed that Christian attitudes underlie this situation.

This raises the question of whether religion can be held responsible for modern developments such as exploitation of the environment. Some might want to argue that religion has little influence compared with secular processes. For instance, Buddhist attitudes are less exploitative than Christian ones to the environment, nevertheless in Buddhist countries such as Thailand the natural forests have been pillaged by large Thai logging companies seeking commercial profits. Indeed it is hard to see how religions can have great effects on reducing pollution and exploitation in the modern secular world. However, it does have its own effects, e.g. the conservation of sacred groves and holy sites, or the ruination of environments by mass pilgrimages such as the Indian Khumb Mela or the Islamic Hajj.

Finally, the authors show that some religions, in particular Christianity, favour economic prudence. One leading analysis of the rise of capitalism (Max Weber's book *The Protestant Ethic and the Spirit of Capitalism*) holds that the garnering of wealth in this world, and especially the saving of money as opposed to the spending of it, arose from the teachings of Protestant Christianity and the attitudes of Protestants. If we accept this view, the real culprit behind modern exploitation of the environment is Christianity.

These and other such themes are explored in the present book.

VERNON REYNOLDS
Oxford

Preface

The book seeks to fill a notable gap in public understanding and in the literature by bringing together two human activities of wide significance: religion and the environment. These have global, national, local and confessional interactions of profound importance not only to ecologists and politicians but to us all. And such interactions are two-way: religion on the environment and the environment on religion. They are considered separately.

The treatment is aimed to be comprehensive and cross-cultural. The environment is the context of all human activity. Soil, climate and landscape profoundly affect human responses irrespective of religious beliefs. But at the same time, religion, and the practices which result from it, deeply mark the environment through the material effects of the basics of human life: birth, marriage, occupation, health and death. Dietary laws and taboos, such as the protection of the cow in Hinduism, the avoidance of unclean animals in Judaism and Islam, and the absence of such rules among most others, influence land uses on a wide scale. Cremations and burials consume resources. Pilgrimages by Christians, Moslems and Hindus involve millions of people annually travelling over long distances, with wide implications for the use of land and route-ways, transport, the disposal of waste, and the risk of disease. Religious buildings, shrines and cemeteries occupy much land, whose sterilization from other uses often makes parts of them effectively nature reserves.

The authors have attempted throughout to avoid bias and approach the subject in a detached and objective way in both presentations and conclusions.

1
Introduction: Religion and the Environment

1.1 Introduction and definitions

A previous study (Reynolds and Tanner 1995) showed that throughout history religions have affected the chances of individuals surviving and their communities expanding or contracting. The present study goes further in showing that religions both affect and are affected by the environments in which their practitioners live.

Religion can be defined as belief in a supernatural power or powers to be obeyed and worshipped and its expression in conduct and ritual. The authors seek to evaluate its interactions with the environment by quoting examples and case studies. They emphasize measurable effects such as religious buildings, group activities, and factors affecting population size and welfare. Some consequences of religious practices are intentional, others unintentional.

Data on religion/environment interactions tend to be most readily available for the larger and more distinctive groups and those whose activities impact on secular politics and the media, notably the great supranational religions; Christianity, Islam, Judaism, Buddhism and the more national Hinduism, even though many of the practices of their adherents do not reflect the views of their theologians or religious authorities.

Environment includes all the natural features of land, water, flora and fauna which support human life and influence its development and character. Some cultures lack a word for 'environment', but may recognize it in other ways. Kiswahili, for instance, makes up for this lack by paying particular attention to personal space in adjectives and enclitics.

In scale the environment ranges from the broad character of a country or continent at one extreme to the minutiae of an individual's living space at the other. In between these extremes are the fields, trees and buildings among which individuals or groups spend most of their lives.

Much recent work on the environment has been concerned with macro-scale considerations because of the increasing concern over the ways in which we have been degrading our habitat. G. P. Marsh, speaking as long ago as 1847 at a meeting of the Vermont Agricultural Society, stated that 'every middle aged man who revisits his birthplace after a few years' absence looks upon another landscape'. His book, the first to be written solely about the environmental damage done by human beings, catalogued these destructive effects, but his approach correlated man with the environment only in general terms.

1.2 The difficulty of recording religious activities

There are few statistics dealing with the religion–environment interface. The main sources are church and synagogue attendance records and information about baptisms, marriages and funerals, but these are only available for some denominations within Christianity. There are no records for how many Moslems attend Friday prayers in even the most important of mosques and no attendance information at all for Buddhism which accords no importance to ceremonies other than funerals. The thousands of religiously independent sub-castes in Hinduism only keep data on their overall numbers.

Most behavioural research is based on relatively short periods of observation and contact with only a few people. Tanner's study of the religion of the Sukuma who perhaps numbered 3 million (1969) was based on a three-year stay in contact with three chiefdoms but probably no more than 12 men were his long-term informants. The number of variables greatly increases in environments such as big cities. Religious behaviour affecting more than single households is difficult to collate. For example, although Hasidic Jews in New York annually live on their fire escapes in commemoration of their people's sojourn in the wilderness, it is hard to measure any environmental effects.

Most faiths have significant religious subdivisions. This is true even in highly centralized faiths such as the Roman Catholics and the Latter-Day Saints as well as in the less numerous Jehovah's Witnesses and Exclusive Brethren. There are several thousand independent Christian churches in South Africa. Hinduism has no central organization and Buddhism has deep divisions between its philosophical and religious extremes. Within Islam, there are behavioural differences between

Sunni and Shi'ites and also between areas where Moslems are and are not in a majority. The behaviour of the single Moslem family on the Isle of Harris in Scotland must differ from that in whole cities such as Mecca where all the population is Moslem.

Nor is there much available information relating religion to such questions as contraception, celibacy, crime or the degree of compliance with dietary laws. It tends to be greatest about extreme events such as the mass suicide at Jonestown or the siege of Waco. Quantitative data about historical events are often unreliable. The reported number of Skopsi men who castrated themselves in nineteenth-century Russia is not necessarily accurate. Figures correlating suicide to religious belief are uncertain where its recording under this heading would affect burial rituals. The Jewish acceptance of suicide as a legitimate alternative to loss of faith, however, makes their figures more accurate.

Large numbers of Christians followed the religious life in the Middle Ages as do many Thai and Burmese Buddhists today. But the assumption of celibacy is not always justified, although the practice is environmentally significant. Unenclosed Roman Catholic priests have sometimes failed to follow it. The martyred last Roman Catholic bishop of Iceland, for instance, had ten children by his mistress. On the other hand, the state pensions of nuns in medieval Europe, particularly Britain, were dependent on their remaining unmarried after the abolition of their religious houses.

Only rough information exists about the numbers of people killed because of their religious faith. The Holocaust is the most recent example, but history gives numerous examples of smaller but still horrendous slaughters. These have often involved strife between Christians and Moslems, as in the Crusades, or internecine strife within these groups as in the Thirty Years War and in the early days of Islam.

There are no accurate data on the extent of compliance with dietary rules, only general indications. Most Moslems fast during the daylight hours of Ramadan, many Roman Catholics used to eat fish on Fridays, and many Hindus are vegetarian. Ethiopian Christians fast extensively, as we know from records of their gastric problems from their restricted diet of flour pancakes and high intake of strong spices.

1.3 The scope of the religion/environment interaction

Religion influences every aspect of individual and collective human life from birth to death. All religions guide and modify human activities in complex ways which vary from one religion to another and within each religion according to such categories as male and female, young and old, clergy and laity, learned and ignorant, or rich and poor.

But the overall impact of religion has been enormous throughout history. It is visible in the ruined tombs and temples in ancient lands such as Egypt, Cambodia, India and Jordan as much as in modern mosques in Baghdad and Marrakesh and cathedrals in Liverpool, New York and the Ivory Coast.

The impact is greatest where religious groups have a lifestyle which is a comprehensive expression of their faith. In medieval Europe the Cistercian monks cleared forests and later changed from farming to sheep rearing. This brought England into prominence in the international wool trade. More recently the Latter-Day Saints and the Amish have extensively modified their environments according to religious prescriptions.

The scope of the religion/environment interaction also depends on the size and homogeneity of the religious group involved. It is greatest in countries where a single religion is strongly predominant as in Japan, Poland, Pakistan or Israel. But it is also great in regions within countries where religious groups, such as Benedictine monks and Hutterites, live in voluntary isolation.

The scope also depends on the stringency with which a religious group controls its members. Although Indian and Sri Lankan monarchs such as Asoka and Kirthi Nissankha Mallawill decreed that animals should not be killed, it is unlikely that such laws were effectively administered throughout their realms. Modern methods of communication have made such controls easier but not absolute. Saudi Arabia is an Islamic state but the extent to which Islamic law can be applied in the oil industry must remain problematic.

1.4 The scale of the religion/environment interaction

The scale of the interaction between religious practices and the environment has both a spatial and a temporal aspect. The spatial aspects vary from the microenvironment of the home which is largely confined to a static enclosure with a limited range of practices, to the macroenvironment of countries and provinces in which religious practices have wide effects on population levels. Even at the microlevel, environmental effects depend on what opportunities the locality provides as well as on the actions of individuals. A single dominant personality working a farm according to religious principles will have a greater environmental effect than another equally committed who lives in a city and who only has a small herb garden behind his house.

Most religions have regional subdivisions, usually based on political boundaries. Christianity is exceptional in having a hierarchy of parish, deanery and diocese all laid out geographically. In some Moslem countries such as Iran and Saudi Arabia, boundaries antedate the adoption of the religion and reflect earlier political and religious subdivisions. Islam at the local level does not have parishes, but towns have many mosques attracting differing groups. Where the members of a faith are a dispersed minority, a different form of religious organization develops as with the circuit system and riders of early Methodism.

Individuals interact with the environment at different scales. Their lives centre in their homes but they also contribute to local and national activities and policies. Domestic religion involves holy pictures on the wall or statues of Buddha. At a local level there are many religious features. Cemeteries contain many memorials. In India, cows wander the streets. In Thailand, spirit shrines are everywhere: in garages, police stations, hotels and on the balconies of apartments (Figure 1.1). At a regional level there are places for the assembly of crowds of worshippers, including cathedrals, mosques and temples as well as buildings and statues commemorating religious figures.

On a large scale most countries have centres of religious significance where they remember important historical figures. Examples are Westminster Abbey in London, Notre Dame in Paris and St Basil's Cathedral in Moscow. Popular religious occasions can bring together huge concentrations of devotees such as Moslem pilgrims on the Hajj, and Hindus at the Kumbh Mela and in Banaras. In the long term such concentrations have a wide impact, using up resources of fuel and food, and generating wastes.

The environmental effect is increased where the group is migratory. This is especially characteristic of desert and semi-desert areas whose inhabitants tend to be nomadic. They range from hunter-gatherers like the pygmies of Zaire or the bushmen of the Kalahari to nomadic pastoralists such as the Bedouin of Arabia and the Tuareg of North Africa. The religious activities of such people are somewhat analogous to those of contemporary modern urban dwellers living in rented accommodation. Many carry protective amulets on their person and practise institutionalized religion mainly for rites of passage and in crises.

The effects of religious activity on the environment also change with time. Most faiths started in the pre-industrial period and each has had to

Figure 1.1 Buddhist shrines in Chiang-Mai, Thailand

adapt itself to circumstances widely different from its origins. The early Christian Church originated in small towns and villages in a mainly Jewish Palestine under the control of Rome. The Prophet Mohammed lived in a pastoral society affected by caravan trading in a part of Arabia never subjected to any imposed imperial rule. The Islamic rules for water and usury are not easily applied in societies provided with electric pumps, reservoirs, computers and electronic banking.

Religions multiply their influence by extending and adapting to new environments. Buddhism originated in India but its main centres today are in tropical South-east Asia and in temperate China and Japan. Christianity has shown a remarkable ability to adapt and influence environments all over the world which have little similarity to rural Palestine or Jerusalem. The more economically developed a society, the more irrelevant such residual influences may appear.

Other societies have changed from oligarchic rival religious states to comprehensive international and inter-faith groupings. The Thirty Years War between Catholic and Protestant states arrested progress in seventeenth-century Europe, but the twentieth century saw the foundation of the World Council of Churches and the worldwide ecumenical movement which have played their part in encouraging overall economic advance. On the other hand, contemporary states, such as those in North America and western Europe, are dominated by major cities which have enclaves of religious exclusivity such as the Jews and the vegetarian Hare Krishna sect. Their rural areas also contain substantial numbers of Christian, Buddhist and Hindu religious communities.

1.5 The effects of religious doctrines

Religious doctrines influence behaviour and this reacts on the environment. We see this most clearly in small circumscribed religious groupings such as sects or monastic communities. They do not always reflect the views or answer the needs of their central religious authority. The Yao of Tanzania are Moslems and also matrilineal. This has prevented them from following the Islamic teaching that inheritance should be through the male line. Islam has no centralized and supervisory institutions to enforce such divine injunctions. Many Sukuma state that they follow religious taboos against eating the meat of certain animals which disappeared generations before. Gandhi's doctrine of non-violence influenced the Chipco movement to preserve trees and led to a massive public resistance to the Narmadi Dam project in southern Gujarat because of the anticipated damage to the surrounding land.

Some faiths require worship to be at specific altars or shines, often in dedicated buildings. These can occupy land, labour and materials for many centuries. But there is always a substantial gap between such requirements and what the believers actually do. Although religious buildings are usually designed for maximum congregations, attendances can be very sparse, even on the designated weekly rest day of some faiths.

Doctrines relating to contraception and abortion have great effects, especially through increasing populations. However, the well-known Roman Catholic rulings against both are followed partly because they coincide with people's interest in children as an economic asset and an insurance against misfortune.

Religious doctrine affects economic dynamism. This generally increases with the degree of independent action allowed to, or encouraged in, the individual as against the group. It has been greatest in the Judaeo-Christian tradition. Where authoritarian concepts and mind control have ruled, they have stifled it. This was seen in Europe in the Dark and Middle Ages and in Islam after the days of Harun al-Rashid. But Islam's early dynamism pushed Middle Eastern trade routes into the Far East well beyond the narrower orbit of European Christian trading. Islam was itself overtaken by more dynamic spiritual and intellectual influences following the Reformation. This was associated with a rapid growth of commercial activity, mainly in Protestant areas, whose effects were seen in the intensification of agriculture, the advance of science and worldwide exploration.

Judaeo-Christian ethical ideas also underlie Communism, which has many of the characteristics of a religion. The intense drive to autarchic self-development by the Soviet government through collectivization following the 1917 Revolution, and its continuation in Communist countries after the Second World War, led to rapid and often careless industrialisation which has left indelible marks on the environment, including the destruction of churches. Collective farms in eastern Europe were so significantly larger than private farms in western Europe that the boundary between them could be seen clearly on satellite photographs.

Tolerance allows a minority religion to have greater impact. While no Jewish or Christian sites are exclusive, some Islamic sites are. Saudi Arabia excludes non-Moslems from the holy city of Mecca.

In highly pluralistic modern urban-industrial societies, the variations in religious practices cause few problems. There is a wide acceptance of Moslems praying on the pavements outside the Regent's Park mosque in west London but not of the public sacrificing of animals for Islamic celebrations (Brise-Barre 1993, 9–12). An overt liberalism does not allow all religious behaviour to be acceptable in all circumstances.

1.6 The intensity of the human/environment interaction

The impact of a religion on the environment is also a function of the degree of commitment of its members both individually and institutionally. This can vary widely between individuals and groups and change greatly with time. Large numbers may nominally belong to a single faith particularly when there is a bureaucratic requirement to record membership, and it is needed when an individual enters hospital or joins the army. But more commonly, individuals are irregular in church attendance, only coming for 'rites of passage' such as baptism and marriage.

Communities have greatest impact where they have shared behaviour patterns such as meeting regularly at a particular place or observing dietary rules. Persecution will induce them to act together in a distinctive way for mutual support. In urban areas the dispersed nature of those belonging to individual churches dilutes their environmental effect. This is especially true with inbred religious communities. The Afro-Arab people in Mombasa, for instance, prefer to marry double first cousins (Tanner 1964, 127–38) in order to keep property within the family following the Islamic inheritance laws. This practice has reduced fertility and increased the chances of having defective children.

Professional dedication will increase the impact of a group. Jains, Jews, Parsis and early Swiss Protestants have become influential through their concentration on business. On the other hand, celibacy, castration and suicide by extreme sects can reduce their growth and even lead to their extinction.

The more people's lives are dominated by their religious practice, the more constant and observable are its effects on the environment. Religious professionals such as priests and rabbis are more focused in consumption and movement than are the laity. The well-organized Mennonites have made clearances in the otherwise uncultivated Gran Chaco of Paraguay even visible to the naked eye from satellite photographs (Sheffield 1983, 117).

A homogeneous religious community will have a more visible effect on the environment than a heterogeneous one, whatever the motivations of its individual members. Purchases and waste products will reflect this. Their composition will vary between vegetarian Cistercians, kosher-keeping Jewish 'kibbutzniks' and meat eating Amish. A contemplative Buddhist monastic community on a minimal vegetarian diet may only impact the environment as a small self-contained nature reserve.

The most intense religion/environment interaction takes place where the religious activities have led to violence. This can lay waste large areas of land and kill both believers and non-believers (Tanner 2000, 101–9).

1.7 The degree of religious centralization

The intensity of the human–environment interaction also falls off with distance from a religious centre because control becomes weaker and external influences stronger. This relationship resembles a series of concentric circles of diminishing control outwards. Tribal faiths, for instance, have intense concern for their own fields, their spirit guardians and the provision of rain, and lessening concern for environments further away where the alien local spirits may be felt to be less favourable. The same applies to geographically nucleated religious groups whose impact on the environment diminishes away from their residential and work centres.

Some societies based on religious principles become highly centralized. Contemporary Saudi Arabia and Iran have religious as well as secular police forces to enforce external conformity to such matters as the veiling of women and hours of prayer. This centralization also operates with small groups. The Zuni Indians of the south-western United States have centuries old hilltop villages and have largely retained their own environmentally related religion, excluding outsiders from seeing their ceremonies. The Hutterites of Canada use a centralized system of administration to avoid disintegration. Many other organizations, such as Buddhist monasteries and Hindu ashrams, need close control at local levels.

In some religions this centralization has been a strong stabilizing factor. The Mormons in Utah have been successful at every level from land use to health and morality. This shows in their demographic statistics, and makes it likely that this policy will continue. The benefits it gives to their membership have discouraged the intrusion of any different religious principles. Similarly the Benedictine and Jesuit religious have created systems of useful centralization which have facilitated their survival.

The centralization of most contemporary religious groups, however, is restricted by secular limitations. Their schools, hospitals and offices have to be run according to the regulations of the societies in which they operate. Islamic charitable trusts, for instance, are administered through a secular Ministry of Waqfs in Egypt.

Centres of important religious faith become focal points for their believers. The Ka'aba in Mecca is the centre of the world for Moslems. Jerusalem has special importance for Jews, Christians and Moslems. The Dome of the Rock is the second holiest shrine in Islam. Tensions can run high as when the Israelis destroyed Moslem buildings near to the

Wailing Wall to provide easier access and a Jewish fundamentalist killed Moslems at worship. It is still a flashpoint in Moslem–Jewish relations.

The centrality of Mecca for Moslems has led to a banning of non-Moslems from the whole area. In spite of this, it has developed into a focus for discontent and this has entailed violence as on the occasion when it was taken over by Shi'ite fundamentalists. Buddhist religious foci, by contrast, seem to have been relatively free from violence.

Any belief in a religious truth seems to create the need to build central places of worship which socially and visually dominate their localities. Since the heavens are seen as above the earth, these buildings tend to be high. This was as true of the ziggurats of ancient Babylonia as of modern temples, churches and mosques. Although the top is not the focus of ritual, it visually represents the effort to approach as near as possible to the divine. Such buildings involve an enormous expenditure in man-power and materials for purely symbolic ends. Their construction can involve a conscious effort to dominate the environment as in Islamic societies where it is augmented five times every day by loudspeakers calling to prayer. In a pluralist society it is a constant encouragement to antagonisms.

Religious buildings in central locations illustrate and encourage particular types of cultural behaviour which are then adopted in peripheral areas. Mosques are egalitarian structures illustrating equality of the prayerful before God. Their interiors are without divisions and can be treated as meeting points. For instance, school groups may pose for photographs as in front of the *qibla* (direction indicator towards Mecca) in the mosque in Cairo's Citadel. On the other hand, Christian tradition has often discouraged social activities in churches by attaching sanctity to their interiors and also because they are often darker than mosques and have more internal divisions imposed by seating, altar rails and rood screens. It must be rare for a school group to pose in front of the altar of a Western cathedral, although newly married couples do so in the Azores.

Centralization in religious institutions can pose health problems. Closed communities can become centres of infection as in medieval pilgrimage centres. The Christian hermits living in the caves above the monastery at Debra Libanos in Ethiopia require an organization of supporters who come regularly to clean the caves to prevent this happening. The forest cave-living Buddhist monks of Sri Lanka also have such a group.

1.8 The degree of religious exclusivity

Some groups are especially exclusive. This may be due to a doctrinal stance as for Orthodox Jews or the Anglo-American sect of Exclusive Brethren, and it is sometimes a defensive response to an external threat of persecution as with the Ba'hai in Iran and some Christian groups in the Middle East.

Such exclusive membership may convey material advantages. Orthodox Jews have special privileges in Israel as do Moslems in Islamic states. It may be associated with a special costume, a defined locality, and receipt of material dividends or welfare. It will involve popular enthusiasm and political pressure which can evoke external hostility. Where governments or social pressures control this situation it can lead to conflicts such as that between Israelis and Palestinian Arabs or to wars such as that between Iraq and Iran.

Some faiths seek to retain exclusivity at the expense of attracting converts. They also try to secure religious homogeneity by political activity. Some Moslem countries discourage tourism and the permanent settlement of non-Moslems. Other groups seek to maintain their position by overbreeding or migration. The Israelis, for instance, are concerned at the low birth rates of non-Orthodox Jews and their encouragement of immigration from Russia owes something to a desire to forestall the Palestinian Moslems from outbreeding them. Some countries deliberately attempt to achieve a religious balance. In Lebanon the distribution of government appointments between faiths was enshrined in their earlier constitution regardless of differential birth rates and actual population totals.

A corollary to the trend towards religious exclusivity is the restriction, or even the exclusion, of minorities. This process has led to tensions, persecution, terrorism, and even war in Israel, Sri Lanka and Lebanon and near-war in Ulster. In Europe before the Second World War, the wealth of Jews provoked envy and hostility which sometimes resulted in pogroms. These features affect the environment through population movements and through making certain areas either exclusive to certain religious groups or uninhabitable as a no man's land, or even, as in Kosovo, sown with landmines.

Religious exclusivity may mask a secular purpose. Individuals may claim membership in a group for material advantages. Some Hindus, for instance, only follow religious practices to benefit from social support for births, marriages, deaths and other such occasions.

1.9 Direct versus indirect effects

There are both direct and indirect environmental effects from religious activities. The former include physical acts such as washing in a particular place or at particular times, using a consecrated building or area, eating a special food, or circumcising young men and women. The latter are the secondary effects provoked by such actions such as increased and redirected traffic flows during festivals or the political effects of restricting access to religious sites.

Some direct effects are unconscious. Many people observe such practices as the wearing of amulets and attendance at shrines. Hindus bathe in the sacred Ganges which their co-religionists have contaminated and contribute to erosion by the deforestation required to obtain fuel for cremating the dead (Figure 1.2).

Indirect factors have effects on the environment which are less clear. The diversion of a road around an Orthodox Jewish community in Jerusalem to conform to religious law may not appear important, but can affect voting balances in the Knesset. A wag complained that the measures imposed on the Likud coalition government by the small National Religious Party could be classified into three: those which were irksome, those which were economically damaging, and those which gave an unfair advantage to the Orthodox minority. Pilgrimages

Figure 1.2 Half-burnt bodies on banks of the Ganges at Banaras

affect transport and trading patterns. Moslems' manners affect domestic architecture in such a way as locating water sources for ritual cleaning and female segregation.

The concentrations of worshippers in religious buildings can spread diseases. In the past these included cholera and plague, but today the curse is AIDS. American Air Force pilots flying to Pakistan on short visits from bases in Britain may have spread Asian influenza to the congregations worshipping in the base chapels. These congregations included many from outside the wire, thus spreading the infection beyond the camp community.

Science, and particularly medicine, has shown correlations between some aspect of religious practice on the one hand and disease and demography on the other. If someone is killed or dies sooner than is necessary because of their religious affiliation or volunteers to be killed or to commit suicide because of their faith, the link is obvious. But there are also unintended effects such as the incidental destruction of neutral property in inter-religious riots. This has occurred throughout history, and recently in India, Indonesia and the Philippines.

The pathways between the religious event and the noted consequences can be tortuous. There is, for instance, a general correlation between regular attendance at religious rituals and lower morbidity, but the causes are complex and long term.

Traditions of regularity, hard work and abstemiousness usually derive from a religious foundation. They are carried over to affect purely secular business or recreational activities. Regular attendance at bingo, for instance, may be organized by the local church, and can bring with it social support and a feeling of fulfilment from spending leisure with co-religionists. The atmosphere is non-judgemental, attracting individuals anxious to assuage feelings of fear and guilt.

Destruction for religious reasons has obvious short-term direct effects. But then the ruins disappear and later generations reuse their materials. Hindu temples have become part of the stonework of subsequent mosques such as the Kutb Minar buildings near Delhi.

Churches and mosques have often lacked sanitation, and have had little protection from the pollution of their surroundings. This has created microenvironments sustaining numbers of non-human species, constituting a network of small nature reserves all over the world. The buildings may also attract tourists and be used for concerts, banquets and exhibitions. In Egypt, the state subsidizes mosques which probably employ over 100 000 men and women. Western cathedrals employ cleaners, ushers, maintenance staff, guides and so on.

Religious buildings may later be converted for use in ways such as trade and tourism never foreseen by their builders. These changes can bring unforeseen effects such as new transport and lodging facilities, but also pollution and disease.

1.10 The effects of religious beliefs on long-term environmental planning

Religious beliefs involve long-term planning for the future needs of the group. The planning of buildings requires projections of the numbers likely to use these facilities as worshippers, students or the sick. While secular building is usually planned for and carried out as a single process, religious building may take decades as money and staff become available.

The religious emphasis on non-material factors reduces concern for the material. This can lead to some neglect in providing for the present world and the interests of future generations. Oriental faiths, such as Buddhism and Hinduism, concentrate on planning people's present behaviour in the light of their expected reincarnation. Some Burmese Buddhists keep account books in order to work out their behavioural debits and credits which might influence their next incarnation.

Whether they plan it or not most religions get involved in property. They collect offerings. They receive bequests of land, houses, money and shares in businesses. This makes them into institutions with long-term interests and ties them to secular stresses. In the nineteenth century four Roman Catholic religious orders, for instance, owned about half of all the land in the four Philippine provinces near to Manila. They had become involved in a wide range of activities which required a more regular income independent of sporadic donations than was possible for self-supporting religious communities. The ownership of this land supported their missionary and educational work. Their business involvement faced them with all the inevitable crises affecting modern peasant societies (Wolf 1971, 50–1) for which their religious principles provided no training or answers. Almost by accident, they were involved directly in environmental issues over the use of their lands against the ideals of their founders. Although they were benign landlords, the rising value of land, changes in the value of cash crops, population growth and the inevitable raising of rents, made them objects of political agitation not only over these issues, but over the wider issue of the conflict between religious and secular power. In a somewhat similar way, the Jesuits had substantial landholdings in Zimbabwe at the time of its independence, which they wisely reduced before it became a subject of agitation.

There is clearly little planning in tribal subsistence cultures whose horizon is limited to the current year. In the absence of a written theology, rituals are unstructured. Traditional ceremonies start when the people arrive. Lengthy ceremonies involve attendees in much coming and going, so that the effect on the environment is more widespread. Through time there is a constant but perhaps largely unconscious process of adjustment to changing social conditions and thence to the environment.

Secular involvements can cause strain when they meet eschatological concerns. These can affect the environment, for instance in the social and economic consequences to the society at large resulting from the building of temples and pagodas. Sometimes the state steps in. The astrological anticipation of worldwide calamities was averted in Burma (Myanmar) by directives from the Ministry of Religious Affairs which initiated the construction of 60 000 nine-tiered pagodas on 9 December 1961 (Spiro 1971, 259).

2
Types of Religions

2.1 The four main types of religion

It is possible to classify religions into four broad types from the point of view of their environmental impact. These are: extinct ancient religions, tribal religions with subsistence economies, the long-established faiths of south and east Asia, and the great monotheistic faiths of Judaism, Christianity and Islam.

The ancient faiths of Egypt, Mesopotamia and central America owed their buildings to individuals about whom there are quasi-historical data. They were monolithic, authoritarian and characterized by vast expenditures of effort on temples and memorials. Egypt and Mesopotamia have long periods of recorded development. Estimates of the resources required to build the pyramid of Cheops at Giza show that it must have exhausted the country more than a major war would have done. Although the underlying desire for large memorials has survived, no other faith has devoted comparable attention to achieving the spiritual survival of its leader.

Others such as Hinduism and Shintoism have no clear initiators. Even Buddhism can be considered in this category since so little is known of the Lord Buddha. His written sayings probably date from three centuries after his death. Jainism had an even longer period after the death of Mahavira before his sayings were recorded. These faiths have seen successions of influential people starting schools of religious thought. On the whole they commit their principles about the spiritual progress of the individual to writing but involve no direct concern for the environment.

Tribal communities seldom have written or clearly defined beliefs. They are usually hunter-gatherers, subsistence farmers or pastoralists and their religions are primarily concerned with their group's survival, the afterlife, and the need to ensure the goodwill of ancestors. They lack the time and resources for complex rituals which do not meet an immediate need. Their devotees tend to abandon their practices when transported into different environments where they have to cope with totally new problems and challenges.

The religions of China and Japan follow the somewhat secular ideas of Confucius and the passivity of Lao Tse. They avoid the interminability of Buddhism, but emphasize the maintenance of a balance with cosmic forces in such activities as the practice of feng-shui (the physical orientation to align with spiritual forces, effectively the terrestrial equivalent of astrology). In the past the state was run through a divinely appointed monarch on astronomical principles, while his subsistence farmer subjects practised their own rituals to ensure survival. Hinduism, Buddhism and Jainism believe in the transmigration of souls through incarnation to an endless succession of afterlives. They see the natural world as illusory and have an overriding mistrust of materialism as obstructing spiritual insights which might limit rather than terminate the uselessness of rebirth.

Judaism, Christianity and Islam are based on ancient sacred texts. They share doctrines of creation which teach an objectively real world in which believers live one life in obedience to God and may aspire to one afterlife.

2.2 The religious practices of ancient societies

The religions of the ancient world tended to be urban-centred. The original location of cities was determined by available natural resources, focal position and the practicalities of site, but within these constraints religion was an important determinant (Adams 1966, 129, 147). There was early urbanization in Egypt, Mesopotamia, Mesoamerica, the central Andes, Indo-China and the Indus valley. The archaeological data show that these sites were large villages or small towns which had a partly religious origin. They all acknowledged allegiance to an external transcendental power.

Luxor, Memphis and Thebes derived their importance from being the centres of worship established by the Egyptian pharaohs. Stonehenge in England was in a settled area and apparently had a religious purpose. Mohenjodaro in the Indus valley had a public bath which is thought to have had ritual as well as practical use. The accumulations of inscribed bones used in foretelling the future found in the earliest Chinese towns were wholly religious in purpose, and could not be the leftovers from early banquets. Some cities such as Kyoto and Angkor have retained religious significance to the present day. There was probably no great pan-continental movement towards building secular cities which did not already have a sanctified nucleus (Wheatley 1971, 92).

Some large constructions were for approaching the gods. Examples are the Tower of Babel, the Luxor–Karnak complex in Egypt, and the ziggurat at Ur, all with remains visible today. Funerary sites included not only the Egyptian pyramids but the underground tombs of the Valley of the Kings and the vast burial pits of Ur and Xian (China). Large areas of houses and markets surrounded these sites. Even the smallest shrine tended to become permanent and popular ones acquired bigger and bigger buildings and survived longer. Some Mesoamerican temple complexes are vast. The so-called Pyramid of the Sun at Teotihuacan which was 213 feet (65 m) high and 720 feet (240 m) square had a volume of 1 300 000 cubic yards (*c.* 1 100 000 cu. m) of sun-dried bricks which had to be carried to the site. Its construction may have taken 10 000 men 20 years (Figure 2.1). However, the remains of houses are often hard to find, indicating that some of these sites were only ceremonial and lacked permanent populations. Monte Alban on a mountain top in Mexico had no permanent water. No community would carve a living site out

Figure 2.1 Aztec Pyramid of the Sun, Teotihuacan

of rock when their essential needs were not near by, so such enormous constructions must have been only seasonally occupied.

Some of these large religious buildings survive to this day. The Cheops pyramid at Giza is enormous but the Mayan one at Cholula in Mexico is even bigger, covering 40 acres (16 ha). The Buddhist stupa at Borobudur in Java is 150 feet (46 m) high. There are some 5000 ancient pagodas within 5 miles (8 km) of a 20 mile (32 km) stretch of the Irrawaddy river at Pagan in Myanmar. The great extent of these reserved areas and the size of the constructions testify to the human instinct for worship and belief in the omnipotence and magnificence of the deity which could not be represented adequately by local or familial shrines.

How many people were taken out of economically productive work to build these constructions and for how long? Some, such as the Buddhist caves in western India, would only allow a limited number of men to work at a time on the rock face so the project must have taken a very long time (Figure 2.2), even if the carvers worked in shifts. On the other hand, religious buildings can be constructed by labourers working on all sides at the same time so that the speed of erection will have depended on housing and feeding them and arranging the regular arrival of materials to the site.

Such constructions were therefore probably built over long periods. The two massive dagobas at Anuradhapura in Sri Lanka which probably together contain 200 million bricks, could have been completed by an

Figure 2.2 Western India. Buddhist cave carved out of rock

annual corvée of 600 men working for 100 days over 50 years (Leach 1959, 15).

Some religious buildings are related to particular rulers, especially as burial chambers. The pyramid of Cheops must have been erected during his reign or at least completed soon after his death since his successors would have had their own afterlives to prepare for.

Some features are smaller than buildings. Stone images are widespread in ancient lands, especially in and around cities. They involved extensive work to crack and remove stones from fields and then to transport them long distances by boat or wheeled transport. Ancient urban centres also had well-developed industries for manufacturing sophisticated ritual objects. The creation of stone and metal images is an advanced skill but would not, for instance, help subsistence farmers to crack and remove stones from their fields by fire, wedges and water (Figure 2.3).

Why did these early people favour such vast constructions? One reason was that the sacred buildings and surrounds were often aimed to be replicas of the wider cosmos, uniting the individual with eternal verities. There was indeed a tendency for kingdoms, capitals, temples and shrines all to be constructed in this way (Isaac 1961–62, 12–17). The geographical arrangement of the pyramids in the Cairo area reflects the configuration of stars in the constellation Orion.

Figure 2.3 Stone images of Buddha in Myanmar

Characteristically, although some of the great buildings from these ancient cities have survived to the present day, the belief systems of the local population have changed beyond recognition. The ancient temples in Egypt and Mesopotamia are centres for tourism today, while near by are the mosques and churches of later faiths.

2.3 Small-scale tribal and traditional religions

Tribal and small traditional communities have limited space and are often under environmental constraints with little opportunity for large constructions of any kind. Religious expression is confined to the limited compass of their habitats. They characteristically use charms and amulets. The Yao in Tunduru, Tanzania, venerate a package of earth from their ancestral lands in Malawi, in somewhat the same way as contemporary Scots view the Stone of Scone. Many modern Europeans perhaps reflect the same view, carrying protective amulets and tattooed symbols.

Anthropological studies have often represented these societies as more distinct and isolated than they are. Each one at its edges merges into its neighbours with whom it shares many ideas and a similar way of life. The hill villages of the Burmese highlands and settlements in the East African cultivation steppe may have remained socially distinct from their neighbours but are broadly similar to them in many ways.

Tribal religions are usually undogmatic and only loosely definable. The active ancestral and other spirits of eastern and southern Africans have always been related to their immediate needs. The people tend to see God as a fall-back answer to any problem. Each family creates its own rituals, following the suggestions rather than the guidance of specialists, who relay the wishes of the spirits and advise on the attitudes of benevolence or malevolence of neighbours and relatives. Every family wishes to expand its landholdings or grazing rights.

The spirit powers in these societies attach to people rather than to land. Thus, items of worship, charms and talismans are carried on the body or kept in the house. The people may have special but unsubstantial spirit houses in the family compound (Figure 2.4) or be connected to animals dedicated to particular ancestors. The most substantial connection is the embargo on collecting only dead wood from spirit groves around springs or rock features in order to protect the environment.

The subdivision of landholdings due to increasing population density is relatively easy but reaches a limit beyond which further subdivision threatens a family's livelihood. The alternative is moving which requires

Figure 2.4 Tanganyika, Usukuma. Ancestor shrines in compound

only the rebuilding of mud and wattle housing which would have had to be done every few years anyway. If this happens the family spirits go with it, often represented by some inherited objects or specially prepared protective amulets.

A typical example of people with such beliefs are the Sukuma of north-western Tanzania. They now number more than 5 million and are treated collectively as a tribe for administrative and political purposes. They are divided into chiefdoms and although each has a different history it is virtually indistinguishable in both language and customs from its neighbours to the south and west. The connecting links between them, however, are too weak to create much unity of belief (Tanner 1958b, 225–36; 1969). In practice Sukuma thinking is limited to the details of their harvests and homesteads within small social units. They only reach out beyond this when there is widespread distress and uncertainty. Since their rainfall is sporadic and showers local, they have rain-making ceremonies when drought affects a wider area than individual fields. For most Sukuma the environment is an abstraction and their language lacks a word for it.

The Bontoc Igorot rice growers of the northern Philippines also exemplify this parochial outlook. They are tied to permanent fields in which there has been considerable investment. Each field has a name and each cluster of fields has its own spirit guardian to be individually propitiated with a shrine. They connect annual rice yields with the particular spirit who controls each valley bottom or hillside. Their concern does not extend to a wider environment as their water supply is from monsoonal rain and does not depend on irrigation water from areas controlled by others.

In Bali, on the other hand, religion has a wider perspective than individual field holdings. This is because the gods and spirits are involved in the equitable distribution of water within a whole watershed. Much the same applies to the pueblo-living Indians of New Mexico who have lived on the same hilltop sites in a semi-desert environment for many centuries. They have a close symbiotic relationship with their gods or spirits relating to the weather and the fertility of their fields. They take care, through their rituals, not to disturb this precarious relationship.

Pastoral peoples have the least identifiable religious connection between their spirit world and the territory over which they graze. For them, the environment is a vast space within which they move their livestock and through which their relationships with the spirit world are channelled. As with swidden (slash and burn) agriculture, annual problems tend to be solved by movement, but the larger problems of drought and flood are ascribed to spirit forces over which they have little control. Some peoples see parts of their environment as the home of gods. The Masai of Tanzania call the prominent volcano in their area 'The mountain of the god Oldonyio Lengai' but have no associated rituals.

Similarly, the Mbuti pygmies of Congo (Zaire) believe that spirits inhabit the surrounding forest in which they move as hunter-gatherers (Turnbull 1961). On the other hand the Nuer, on their flat floodplain in the southern Sudan, believe that their gods are nowhere in particular. In none of these cultures does the spirit world have any discernible influence on anything more than the ground on which they happen to be at any particular time.

2.4 Chinese and Japanese religions

Chinese and Japanese religions differ from others centred on sacred books in having ideographic rather than phonetic systems of writing. To read a newspaper may require learning as many as 3000 characters and to read religious books requires many more. Therefore, while these religions are well recorded in the writings of Mencius, Lao-Tse, Confucius and the Indian Buddhist sutras translated into Chinese, they can only be read by small numbers of highly literate scholars. Indeed it has been commented that the key terms in these works are never defined perhaps because the need for this never came up. In consequence, the religions of these peoples have been largely non-literate and confined to a relatively small range of theological ideas.

The government of China for several millennia has been centralized. It had a sacrificial system which began even before its dynastic rule in 2205 BC and only ended when the Manchus were deposed in AD 1911. Shang-Ti was worshipped as the creator God. Every year the Chinese emperor conducted a 'border sacrifice' on behalf of the nation. Confucius compared the comprehension of its ritual to the efficient ruling of the Chinese empire. In the fifteenth century AD this sacrifice was moved to the Heavenly Temple in the southern part of Beijing in which there was no idol.

In general Chinese religion tends to be a blend of Buddhism, Taoism and Confucianism, including also a belief in feng-shui. There are over 100 sacred mountains in China, but among these are 9 formal sacred mountains, 4 Buddhist and 5 Taoist. On the Buddhist mountains the visitor is essentially an observer but on the Taoist mountains he becomes a part of nature and the visit is a total experience (Palmer 1998, 28).

Locally, it has shown extreme diversity with individuals and groups performing their rituals combined with a relative indifference to doctrine and a minimal role for the professional clergy (Deglopper 1974, 43–5). Each of the innumerable farming villages would have worshipped the spirits in every street, neighbourhood and field. They also worshipped ancestors with regular visits to the family graves.

Agricultural life in China was always precarious due to natural hazards and political disturbances. The worst hazards have been floods from the Yellow or Yangtze rivers breaking their banks and periodic droughts, which must have made an extra-mundane refuge appear attractive. The Buddhist concept of reincarnation may have appealed to people leading miserable lives, especially if they misunderstood that such reincarnation was not necessarily into another human form.

Buddhism came to China after innumerable local cults had become part of the people's everyday life and thought. It brought a monastic system from other centres of the faith. The *bodhisattvas* (people who through countless lives have been able to leave the world of birth, death and rebirth) clearly advocated care for nature. Compassion led to keeping the land around Buddhist temples and monasteries as sanctuaries for wildlife because religious tradition forbade killing in such places (Palmer 1998, 24). Monasteries often featured a pool for the release of living creatures near the main temple gate and stables for the care of

Figure 2.5 Japan. Deer in temple grounds in Nara

rescued animals, where the sexes were kept separate (Welch 1967, 378) (Figure 2.5).

Taoism sees nature as having an inherent balance but, unlike Confucianism, does not believe in its control. It has four principles (Palmer 1998, 22–4):

1. *Tao Te Ching*, emphasizing the importance of the earth and that we should obey its movements.
2. *Yin and Yang*, believing that energy comes from the balance of the two principles. In practice this means making full use of nature by observing and following its ways.
3. Remembering the limits of nature's sustaining power and that it will punish those who overuse it.
4. Nature's quality to be judged by the number of plant and animal species.

Taoism provided magical and material safeguards for prolonging human life and was closely linked to medicine, defence, charity and welfare. Its founders were from the lower strata of Chinese society and it has flourished where its religious leaders have practised magic and emphasized millennial possibilities. This has caused it to be periodically involved in political uprisings, and rebellions such as such as the Yellow Turbans, and the Taiping and Boxer rebellions who opposed the premise that the Emperor as the Son of Heaven ruled from Beijing (Eickhorn 1959, 385–6). Nevertheless, its most obvious characteristic affecting the environment is the passivity it induced.

Confucianism is more a code of manners than a religion. It judges belief and disbelief by their effect on human conduct and balances them with scepticism. Its Five Cardinal Principles, benevolence, duty, manners, wisdom and trustworthiness, have no religious backing except in following the Way of Heaven (Graham 1959, 365–8). It may have come to prominence during a prolonged period of peace and economic well-being which made fighting seem irrational and unnecessary (Wilkinson 1980, 92). It requires reverence for ancestors, natural phenomena and men of the past rather than for personal gods. There were always a plethora of cults from that of each individual family to that of the Emperor as the Son of Heaven (Maspero 1950). Overall it has a functional rather than a sectarian approach to the questions of identity involving ancestor worship, divination and geomancy (Strikman 1982, 55). But it is an elitist system of thinking and behaviour which can have had little appeal to the majority of peasants. Chinese religion has no word equivalent to the Western 'nature'. In such concepts as the Tao, the Ten Thousand Things, Yin and Yang and the Five Elements, human beings come from and grow in nature and return to nature.

In Japan, although there never was a single cult, Shintoism was centralized and had a pantheon of gods supposedly totalling 8 million (Piggott 1969, 42). It followed a more Taoist path based on native and indigenous spirits connected with unusual objects. Believers expressed praise and gratitude to ensure their continuing benignity by praise and gratitude (Bownas 1959, 348–9). They approach problems in a matter-of-fact way. Human beings are part of the flow of things and have no particular possibilities over and above other kinds of beings (Bucking 1994, 164). Shinto became the religion of Japanese patriotism but remains close to the traditional religions which, as with the Chinese, are local rather than universal.

The ordinary Chinese and Japanese people do not, unless subjected to a process of indoctrination, readily express themselves through a dialectic or an ideology built on sophisticated premises or logical inference (Noss 1959, 399). Their popular religion is anti-intellectualist and pragmatic (Meslin 1972). Any visit to temples and cemeteries in Taipei, Tokyo and Manila confirms this popular approach, as seen in family meals on graves and the burning of paper money to share success with the ancestors.

In both China and Japan the age-old gap between semi-official national politically oriented cults on the one hand and 'folk religion' on the other leaves little room for environmental concepts, both because of a lack of emphasis on the topic at either level and because most environmental questions fall in scale between the two. The central government is concerned with large-scale economic planning, the people with the details of their daily lives.

2.5 Hinduism, Buddhism, Jainism, Sikhism and Zoroastrianism

Hinduism

Hinduism, Buddhism, Jainism and Sikhism all originated in the Indian subcontinent with the first being the parent of the others.

Hinduism is the third most widely followed religion in the world representing around 14% of its population, behind only Christianity (34%) and Islam (19%) (*Encyclopaedia Britannica* 1999). Little is known of its origins and it is best understood from its writings. The first were the Vedas (*c.* 800–600 BC) which originated with Aryan invaders in India. Scholars later developed these into the more philosophical and mystical Upanishads, and the romantic and poetic Mahabharata and Bhagavadgita. The main common thread in Hinduism is the belief in the transmigration of souls and in a great variety of godlike figures from which individuals and groups can choose who, where and how they will worship. The principles of human action derive from the concepts of *dharma* and *karma*. Dharma is the cosmic order which includes both natural and moral principles governing people's behaviour within their caste. Karma is the view that a person's actions in one incarnation determine his fate in the next. A general effect of these beliefs is to encourage an acceptance of fate and to discourage originality of thought or action. Apart from these there are few commonly agreed beliefs.

Hinduism today has no corporate identity or centralizing institutions and few commonly agreed ideas. It has indeed found it impossible to think in terms of orthodoxy, and has allowed polytheism to flourish with its main gods and goddesses and innumerable local deities. The absence of centralization, combined with the versatility of caste (Milner 1994, 99–109) has made it a responsive and adaptable part of contemporary Indian life. The same lack of centralization also appears in the Balinese system of separate temples controlling the water supply to rice fields within each river's watershed. The caste system historically was a complicated loose framework with no overall simplicity on to which British administrators and researchers imposed the idea of rigidity. Caste has become more important as a consequence of constitutional provisions since 1948 and the recognition that voting can change social disadvantages. A candidate for election can appeal to and gain the support of caste members by assuring them that their exclusive needs within a religious framework will have special attention.

The particular importance of Hinduism for any study of the environment lies in the practices which derive from its amorphous and polytheistic character. Despite, or perhaps because of, this character, it has been able to generate innumerable cults and sects without ever having built any central institutions with an interest in making and maintaining orthodoxy.

There are four main castes: Brahmans (priests), Kshatriyas (warriors), Vaishyas (originally peasants but later merchants) and Shudras (serfs). The fifth group is the Outcastes, or Harijans, beloved of God according to Mahatma Gandhi. Castes are based solely on inheritance and ascription so that a person's social status is unchangeable. It continues to exist among converts to Islam and Roman Catholicism in southern India and as well as among the Goans of the former Portuguese enclave. The main structure has been divided into a large number of sub-castes, each with its own social network and required behaviour. As society evolves more occupations, new ones are added.

The castes are geographically intermingled and interdependent so that they do not inhabit separate areas as do, for instance, the Amish and Hutterites of the United States and Canada, although the very numerous Outcastes generally live on the outskirts of villages with their own wells. Hindu society cannot do without them as they dispose of rubbish, including human and animal bodies. Although social and economic change has blurred vertical distinctions between castes, the absence of any effective social security system makes their horizontal social importance more significant, particularly in politics.

The Outcastes still have inferior status despite political representation, constitutional guarantees and reserved opportunities in education and the civil service. They suffer considerable hostility and violence which has led them to develop separate professional structures and caused many thousands in western India to convert to Buddhism.

Obedience to caste is deeply embedded in their consciousness. In the early nineteenth century a British traveller wrote, 'When my father returned to the boat after failing to persuade the poor dying creatures to eat our prepared food, one and all made the same reply, "Sir if you will give me food cooked by a Brahmin I will eat, if not I must die"' (Sattin 1986, 10), Similar beliefs continue and were reported during the 1943 Bengal famine.

The most sacred city is Banaras on the Ganges. Pilgrims bathe in the river which is lined with ghats (flights of stairs), leading to 1450 temples. The chief buildings are the Golden Temple of Siva, the Nepalese Temple and the Monkey Temple, and there is the Hindu university and Brahman schools. Believers cremate bodies on the Burning Ghat. However, the city also contains hundreds of mosques, the chief of which is the Mosque of Aurungzebe.

Hinduism has a close symbolic as well as functional relationship with animals, plants and trees. The cow is sacred as a deity and as the incarnation of Vishnu. Because of its association with Krishna and the milkmaids, the cow wanders the streets of most cities and helps itself to food in grain shops whose owners do not protest (Figure 2.6). In a primarily agricultural society, cows are valuable since they provide both food and fuel (Pannikar 1979, 286).

Other animals also have quasi-divine status. Hanuman is the monkey god who came to the help of Rama with his army by building a bridge of rocks between India and Sri Lanka. Ganesh is the elephant god of wealth and prosperity celebrated in a contemporary Bombay festival with an inflated model on the beach. Other gods and goddesses are associated with animals and birds: Siva with the bull, Durga with the lion,

Figure 2.6 Hindu India. Cow in streets of Jaipur

Kartikeya and Sarasvati riding on a peacock and swan respectively. Other sacred animals include the goat, buffalo, tiger, snake, rat and owl. Sometimes temples are dedicated to particular animals as the rat at Karni Ma in Bikanir.

Vegetarianism has a long history in India since it is mentioned in fourth-century BC descriptions of Megasthenes. It is common in contemporary society although often associated with higher social status where the opportunities for a varied diet are greater. This practice and that of non-violence are connected to the ancient and prolonged influence of Buddhism and Jainism.

The belief in the transmigration of souls makes humans and animals interchangeable when the circumstances of life demand it. Therefore, humans do not necessarily have an unchallenged superiority unless they have ultimate knowledge. They are seen as equal to animals in intelligence because both are ignorant in knowing only the sensual world (Choudhury 1994, 74–5).

Hindus are still predominantly subsistence farmers. They see the surrounding forests much as Christians view the wilderness and have the same wish to protect it. The 'Chipko-hug' is a tree movement started in 1974 by Garwhali women which stopped a company cutting trees for the manufacture of sports goods.

Hinduism deifies certain plants and reserves special days in the calendar for their worship. The bodhi (*Ficus religiosus*) under which Buddha achieved enlightenment is protected in many villages as well as the original tree at Bodh Gaya near Banaras. Ramakrishna, a nineteenth-century saint, had his primary vision under a Bengal variety. The hibiscus is associated with Kali and Durga and the lotus with Lakshmi. The use of plants is a major part of the Ayurvedic medical system and the Atharvaveda mentions 110 plants with medical uses.

Hinduism perhaps appears most clearly as a cohesive social and religious force in relation to Islam. The two faiths have intermingled for centuries in India. Their members have regularly rioted and occasionally waged wars. There were riots at the time of partition of India in 1947. In one riot alone, 3000 dead were removed from the streets of Calcutta. Since India and Pakistan have been separate nations they have fought three wars. A recent flashpoint was the attempt to build a temple at Ayodha on the site of a mosque because it was the alleged birthplace of Krishna.

The influence of Hinduism on the environment is therefore mainly notable for its lack of central administrative controls, the segregated living of different castes, the widespread observance of a vegetarian diet, the ubiquity and wide tolerance of cows, the religious use of the Ganges river, especially at Banaras, and overall the separation but interdependence of castes.

Buddhism

Buddhism began with Gautama's revelation of the 'Eightfold Way' and has followed Hindu beliefs in transmigration and reincarnation. It has two main subdivisions: Theravada and Mahayana. The former derives largely from the work of the monk-king Asoka and is found in those areas which remained in the faith after the Moslem invasion: Thailand, Burma and Sri Lanka. It reflects the earliest form of the faith, being relatively pure, conservative and near to Gautama's original teachings. Mahayana Buddhism also dates from the time of Asoka. It has a different canon which has never been codified. Its sacred books are written in Sanskrit and it has become the religion of a vast number of societies across central Asia, Tibet, Mongolia, Korea, China, Japan, Vietnam, Nepal, Java and Sumatra. It has transformed the faith in accordance with the different backgrounds of these lands, although its adherents claim originality and the sanction of the founder for their views and practices. The main practical differences from Theravada are in the deification of Gautama himself, the replacement of full Enlightenment with the concept of Compassion for humanity, and the much wider use of images to supplant Gautama's original atheism (Bentley-Taylor 1950, 129–31).

It is hard to see in Buddhism anything remotely connected with concern for the environment (Harris 1991, 101–14). It is essentially a world-denying religion, even questioning the world's reality. The British Library's standing exhibition has a copy of the Perfection of Wisdom Sutra from Tunhuang in north-west China, originally produced in India in the first century AD. The explanatory label writes, 'Its philosophy may be summarised as – the nature of all phenomena (*dharma*) is non-nature and their non-nature is their nature. Their one characteristic is having a non-characteristic.' This, together with the earning of calculable merit, and the worship of particular personalities, was the basis of most Buddhist practices wherever situated and at whatever level of sophistication (Robinson 1959, 321–9).

The Buddha taught detachment from all the things of this world and his followers were expected to detach themselves even from the concept of 'self' if they wished to leave the Wheel of Existence after endless reincarnations. There is nothing in such doctrines to uphold an environmental ethic since even world destruction would appear to aid rather than defer this process of detachment from the self. The statues in some temples of emaciated saints show the unimportance of even a minimally satisfying standard of living. Orthodoxy would view both environmental care and environmental destruction with equal indifference.

Buddhism, unlike Hinduism, allows no rituals for personal protection and satisfaction. Its dominant philosophical approach to human needs has caused the overwhelming majority of nominal Buddhists to move away from the strict early doctrinal stand of their faith. Despite its concept of endless rebirths they have concluded that pleasure need not involve suffering, and that nirvana is not extinction. They now tend to believe that individuals can through their actions acquire merit to improve, rather than extinguish, their karma and so obtain salvation. In addition they have developed a religion of ritual protection since Buddhism provides nothing beyond meditation. They assume that they can obtain relief from all the worries of life from specific ritual acts which either earn merit or enlist the assistance of supernatural beings and power (Spiro 1967, 140).

Buddhism has always inspired the building of pagodas, often large, and has given much effort to their maintenance and beautification. Cambodia provides some outstanding examples. King Jayavarman VII built more in the four decades of his rule (AD 1181–1218) than all his predecessors put together. The details are documented. The constructions included 121 resthouses for pilgrims visiting Angkor Wat (Coedes 1940, 247–9), as well as a 10-mile long wall with its 100-yard wide moat round the ceremonial enclave (Mus 1936, 57–68). King Parakrama Bahi 1, who reigned 1164–97, is credited with building 101 pagodas, 476 Buddha statues and 300 image chambers. Both these monarchs also constructed or repaired an even larger quantity of public works. This testifies to enormous personal energy and the existence of a complex and efficient administrative system.

The faith is also concerned with all creatures, even microbes, as the possible repositories of some transmigrating essence, not the soul. It characteristically avoids any interference with the natural processes of the animal kingdom. In Tibet they will not even kill personal parasites. The most characteristic Buddhist environments are the monasteries whose monks are prevented from working the land because of the possibility of damaging living creatures. In economic terms these men, and to a lesser extent the nuns, are parasitic on society, since beyond meditation, they are only concerned with their own processes of detachment from human involvement and have no pastoral role in the surrounding communities except to perform funeral rituals. They are not

Figure 2.7 Myanmar. Buddhist monks on their morning begging round

involved in other rites of passage except as the beneficiaries of gifts. They must depend for their survival on their daily morning begging round (Figure 2.7).

The monasteries are in practice nature reserves for birds, monkeys and dogs (Figures 2.8 and 2.9). They spare the fish in the nearby rivers which congregate near the landing stages. Many monastery compounds have baskets of fish, frogs and birds which people can buy and release to acquire merit towards a better reincarnation. They can also gain merit from building pagodas and in decorating both them and statues of the Buddha with gold leaf.

Modern westernized Buddhism romanticizes the past, but implicitly makes a doctrinal shift away from fatalistic earlier doctrines. The key-notes to Buddhism's environmental impact are probably passivity and conservation. The passivity opens the way to authoritarian and immobile government and the conservation leads to the preservation of vegetation and animals within traditional patterns. But today Buddhists are increasingly adapting to the commercial dynamism originating in other societies.

Jainism

Jainism was founded by Mahavira in about 600 BC, and can be seen as an extreme sect within Hinduism or even as a schism. Today it has 3 million adherents but these only form 2 per cent of the Rajasthan

Figure 2.8 Hindu India. Monkeys in temple

Figure 2.9 Bangkok, Thailand. Dogs in temple grounds

population where they are concentrated round their temple complex on Mount Abu. It has quite distinct doctrines and the social characteristics which result from them. Barth (1992, 147) described it as a 'disordered open system'. In AD 52 the 'sky clad Digambaras' living in the warmer south who favoured nudism divided from the 'white clad Svetambaras' in the colder north who favoured clothes, a severely practical rather than theological distinction.

Jain prescriptions are so extreme that only a small proportion of its adherents, termed 'renouncers', attempt to follow them. These live in clusters because the practice of Jainism requires proximity to a temple. They build monumental stupas in commemoration of their saints.

The highest duty of a Jain is not to kill any living being. Most would interpret this merely as a moral injunction to be a vegetarian, but the religious teaching goes much further. Its vegetarianism is extreme and there is also a complex list of requirements of varying severity from the Five Lesser Constraints to the Twelve Constraints and the Fourteen Disciplines. These add limitations on clothing, travel, furniture design, water use and sexual relations. Believers wear veils over their mouths, dust seats before they sit down, and are reluctant to walk at night or to use naked lights in order to prevent harm to insects. This renunciation of violence even extends to arguing. They avoid farming in case the plough damages the soil and kills animals and insects there. They do not boil drinking water for fear of killing bacteria which might carry the soul of another living thing. They make charitable contributions to provide sanctuaries for old, sick and stray animals, even for insects, where they keep them until they die a natural death. Jainism is probably unique among religions in not putting human survival at the centre of its practices, if not of its beliefs. It is the only faith that gives humans no priority over any other living things, even a plant, putting all on a par.

However, Jains do divide living beings according to the number of sense organs they possess. The highest is *Panchendriya*, which includes the mammals who possess all five: taste, smell, hearing, sight and touch. The lowest plants are *Ekendriya* which possess only one. A sacred Jain text refers to the torture undergone by trees in being cut down and cut up. It is easy to see that such beliefs if practised would impose an intolerable, indeed impossible, burden on anyone wanting to continue living. It has even been suggested that the true Jain who has come to the end of his useful life, should allow himself to starve to death (Basham 1964, 261–6). Carried to a logical conclusion this would lead to the extinction of Jain communities.

Since they have to survive, they eat the rice and vegetables cultivated by others who will then bear the main burden of the harm they have done. They use their annual temple fairs as a propaganda vehicle for vegetarianism on its own health merits rather than as a religious requirement involving conversion. Their reluctance to risk harming living creatures has led many into commerce as the only religiously safe occupation. Overall, they are a property-owning plutocracy with a lavish public life in which moral, social and economic standing is intimately interconnected (Laidlaw 1995, 87–147). Their success with money lending accounts both for their high average wealth and their unpopularity (Jacobi 1914, 465–74).

The Jains, by seeing their lives as part of an endless transmigration, are virtually ignoring the consequences of such extreme asceticism. The demanding nature of their beliefs and their geographical dispersion have always restricted their numbers so that they have had relatively little group impact on the environment, except for a limited effect on preserving flora, fauna and some buildings.

Sikhism

Sikhism was founded in the Punjab by Guru Nanak Shah (AD 1469–1538) in the late fifteenth century, and the great majority of believers still live in that province and neighbouring areas, although they now have large communities in the USA, Canada and Britain. Today they number about 22.5 million worldwide. They are disciples of the ten gurus beginning with Nanak Shah and ending with Gobindh Singh (AD 1666–1708).

The faith contains many unique postulates and principles which are quite different from both Hinduism and Islam. In some ways they approach monotheism. They believe in a single formless god with many names that can be known through meditation. They do not deny the existence of the countless Hindu deities, but only allow worship of one supreme god. They reject the caste system but agree with Hinduism in accepting reincarnation.

They constitute about 66 per cent of the population of the Indian state of Punjab and so its character to a large extent reflects their beliefs and practices. It has a productive, increasingly commercial agriculture, and a diversity of industries among which textiles and transport equipment are notable. It has the highest per capita income in India. With less than 2 per cent of the total area it produces more than 10 per cent of the grain and more wheat than all the other states combined. Commercial fruit and vegetables are also important. All these have benefited greatly from the 'green revolution' in the country over the past half-century.

In practice Sikhs are hard-working efficient farmers and traders, often filling official positions as soldiers and administrators in India and elsewhere. Strict Sikhs carry the 'five k's': *kesa* (long hair which is never cut), *kanga* (comb), *kacha* (short pants), *kara* (metal bracelets) and *kirpan* (ceremonial dagger).

Sikhs meet in temples of which the chief is the Golden Temple in Amritsar. Of recent years the community has been split between liberal and conservative wings over whether or not to use tables and chairs in the temples. Since 1940 they have sought a separate state and it was Indira Gandhi's Sikh guards who assassinated her in 1984 for ordering the army to attack the Golden Temple in Amritsar which led to the deaths of more than 450 of their fellow believers. Overall Sikhism has given rise to progressive, hard-working communities with an efficient use of land and a representation in official positions and in the professions beyond what their numbers would suggest.

Zoroastrianism

Zoroastrianism is the ancient pre-Islamic religion of Iran that survives there in isolated areas and more prosperously in India where the descendants of Zoroastrian immigrants, who came to escape Islamic persecution, are known as Parsis. Founded by Zoroaster in the sixth century BC, the faith contains both monotheistic and dualistic features. Zoroaster saw evil as having originated from a conflict between the two sons of the one god Ahura Mazda, the 'Wise One'. One began evil, the other good. He taught that the world would end in a great conflagration. Until then the dead would be separated into those destined for heaven and hell.

Parsis originally settled in Gujarat but live today mainly around Bombay, Karachi and Bangalore, though many have migrated to other parts of India and elsewhere. Although not a caste, they are a well-defined community. By the nineteenth century they were prosperous and renowned for their liberality and support for education, including that of women. But their numbers are steeply declining: from 91 000 in India in 1971 to 71 000 in 1981.

Their impact on the environment is relatively small and comes mainly from their demand for materials surrounding birth, marriage, annual feasts and death. The young male Parsi experiences a ceremony of purification in cow urine at the age of 7 and is invested with a woven 72–strand girdle. The priest's benediction includes throwing on his head slices of fruit, seeds, perfumes and spices. Parsis are strictly monogamous. This reduces the birth rate, especially since many marry outside the community and the career-mindedness of highly educated Parsi women makes for late marriages. The wedding ceremony involves binding the couple with cords and throwing rice on them. Eight days later, the newlyweds entertain relatives and friends to a vegetable banquet. Parsis hold annual feasts in honour of creation, crop planting, mines and other environmental features. Their biggest impact on the environment is probably through their funerals which are aimed to 'avoid polluting nature with rotting flesh', discussed further in Chapter 10.

2.6 Judaism, Christianity and Islam

Finally there are the three 'book' religions of Judaism, Christianity and Islam which seek to relate the Old Testament, the Bible and the Koran respectively to the modern world. All look back to Abraham as a common founding patriarch and see human life as preparation for a single afterlife and the environment as a gift from God for which they are responsible. Their sense of responsibility goes beyond viewing the environment as a neutral entity. They see nature as a means by which humans come into relationship with God. It incarnates the word of the creator just as art incorporates the word of the artist, and it offers entry into a dialogue with him. Humans can be his partners in helping to shape it further. He will even increase the yield of nature to those who repent their faults and undergo spiritual revival. At the same time, God is not only the creator of nature but has allowed himself to be subject to its laws. These ideas have shaped and fuelled civilization in all areas followed by these faiths.

Judaism

Judaism perceives the environment as a benevolent gift from the one true God for which humans have stewardship. The Jews received the Promised Land after the Exodus and the giving of the Law. Their occupation of the land carried with it responsibilities relating to a comprehensive range of social needs governed by moral, civil and ceremonial laws centred in a system of temple worship. These were based on an understanding that fallen man's relation with God could only be re-established through blood sacrifice.

They continue to see the human occupation of the natural environment as reflecting and symbolizing God's plans for this world and the next. The rules for eating which are such a conspicuous feature of Orthodox Jewish life are a perpetual reminder and symbol of this (Douglas 1966).

The prosperity of the land depended on the people's obedience to the divine covenant 'to love the Lord your God and to serve him with all your heart and with all your soul. That I will give you the rain of your land in season . . . that thou mayest gather in the corn and thou shalt eat and be full' (Deuteronomy 11: 12–17). In Old Testament times this required fallowing the land every seventh year by analogy with the Sabbath rest day (Exodus 23: 11). Modern Israel observes the civil aspects of these principles. Only a handful of agricultural collectives observe the 'Sabbath of the land' in its biblical and rabbinic sense, but the principle has influenced the country's scientists to question intensive agriculture (Solomon 1994, 111–12).

The destruction of European ghettos by the Nazis led to the dispersion of Jewish populations and a move of their environmental influence to Israel and a few large Western cities, particularly New York, where their communal activities are confined largely to home and synagogue. The state of Israel is a unique case with agricultural settlement patterns which contrast markedly with those of their Arab neighbours.

The biblical injunction to be fruitful and multiply within a set of behavioural patterns, in which restraint is related to menstruation, has enabled Judaism to maintain its numbers despite persistent persecution.

Christianity

Christianity derives from Judaism and likewise sees the natural world as a creation of a benevolent and rational God who wills that the universe and its inhabitants should exist and prosper. It believes that a worldwide fellowship of believers has replaced the theocratic state in the divine plan for the world. The primary focus is on the salvation of individual souls and the ennoblement of their mutual relations.

The Christian attitude to the environment stresses the dependence of communal action and economic progress on the moral qualities of individuals. These follow from the observance of divine injunctions.

Much biblical imagery is agricultural and pastoral within a village setting, although Paul's letters were mainly to urban communities. The faith has proved adaptable to every variety of setting, but its influence over the environment is most clearly seen where Christians have been concentrated in enclosed communities such as monasteries and isolated colonies such as the Mennonites, Amish and Hutterites. Monks and nuns practise a physical self-denial and single-sex communal living which is largely absent from both Judaism and Islam. Monasticism has often seen extreme asceticism as the ideal for a person's spiritual development. The urban believer's most obvious environmental influences are in church buildings.

The biblical injunction to be fruitful and multiply, and the moral objections to contraception shared with Islam and Judaism, have favoured population growth. But some Church doctrines may have inhibited the growth of Christian communities. The pursuit of sanctity may have restrained sexual activity because of the latter's association with sinfulness. Catholic Christianity not only forbade artificial contraception but also imposed wide-ranging restrictions on sexual intercourse, forbidding it on saints' days, during Lent, before taking communion and enjoining long periods of abstention as a penance.

Some Christian groups deliberately include environmental protection as part of the biblical way of life. The Amish, for instance, reject petrol engines and cultivate with horses which do not disturb the soil as much as tractors. Since their interpretation of the Bible does not allow the use of profit for the purchase of consumer goods, they put much money back into farming and into assisting other members of their communities to get started. The success of their agriculture and their large families, however, has meant that they are outgrowing their existing communal areas. They have had to purchase new farms beyond easy visiting range with horse-drawn transport, necessitating an expansion of their social organizations.

The Hutterites have a much more communal life than the Amish. Families do not have their own houses, but their success as farmers has not led to the same problems of expansion since their farms are in low population density areas of Canada and Montana. They subsidize communes to split when they become too large for communal living.

Different branches of Christianity have noticeably different environmental effects. One example is the difference between Protestant and Catholic Europe seen, for instance, in travelling from the Netherlands into northern Belgium, from Basle to Belfort, or between the northern and southern parts of Ireland. Despite geographical proximity and close environmental similarity, their appearances and land uses differ significantly. Crossing the border out of Ulster one leaves behind the relatively neat, well-ordered but somewhat austere landscape of small fields, grey stone villages and simple churches into a wider, more unkempt seemingly deeper green and lusher landscape with an absence of barbed wire and occasional Italianate churches somewhat reminiscent of Spain or Portugal.

Islam

Islam arose as a reaction to the moral decline of a tribal and pastoral society. The Koran records Mohammed's earlier visions of moral regeneration and later ones of communal organization. Early Moslem practice reflected the behaviour of a migratory people in an arid land. Their lives involved seasonal transhumance, trading by camel caravan and living in towns whose desert surroundings and wide separation make them somewhat reminiscent of seaports in character. Mecca and Medina provided the examples and stimulus.

The early shrine at Mecca was under the control of the Quaraish clan, whose growth in political and economic power correlated with the increasing importance of the Ka'aba, the religious core of the Moslem world. It was in the clan's interests to free a sacred area with a diverse and transient population from tribal rivalries. The shrine was in a site which can have had few attractions for a pastoral people and none at all for farmers. Its success was no doubt helped by its relevance to political, economic and social objectives, and was developed by those who would have been able to organize the necessary workforce for its building and maintenance. The Quaraish tried to extend the sacred area to provide them with better trading opportunities but as this was invading the power of other groups, they did not succeed. Mohammed's subsequent endorsement of the reasonableness of allowing pilgrims to trade helped the development of the city.

Mohammed brought a stable system of government, administering rules believed to be inspired by a God who was not linked to any particular traditional society. The surrounding environment was free from external threats as it came at a time in history when both Persian and Byzantine empires were moribund. This partly accounts for the rapid spread of Islam in the seventh and eighth centuries.

Mohammed's rules were straightforward and not too difficult to follow: an affirmation of faith, a once in a lifetime pilgrimage to Mecca, fasting for the daylight hours for one lunar month a year (easier where there is little seasonal variation in daylight duration), charity to the poor and needy, and regular prayers which could be carried out anywhere without priestly intermediaries. Moslems can say them in shop, library, market place or even in an aircraft in flight. But they sometimes ignore some of these injunctions, particularly as Islam has no supervisory institutions.

The practices of Islam were designed for believers who belonged to a geographically mobile Arab society, and wished to differentiate themselves from Christians and Jews. The Moslem state has always combined moral with civil authority. This, reinforced by the practical demands facing early believers, justified the rejection of a permanent priesthood. This made a permanent priesthood both undesirable and impracticable. Moslems are thus able to approach God directly without the use of intermediaries. But Islam lacks Christianity's intense emphasis on overcoming personal sins. Its view of God as being arbitrary and unknowable tends to breed fatalism which undermines initiative and hinders any impulsion to work at improving circumstances.

The Moslem laws which have had the most identifiable effect on the environment are probably those enjoining the division of heritable property and controlling land tenure. The almost inevitable result of the former has been the subdivision of land and trees into smaller and smaller units unable to feed families, and the separation of holdings so that time is lost in moving between them. Ludicrous situations result such as several people owning a single olive tree or date palm.

The system of land tenure poses problems because of the insecurity it involves. In the irrigated areas of southern Iraq, for example, the pre-1918 Ottoman land system has largely survived to the present day. Most of the cultivated land is held by landlords under leasehold systems which, opposite to Western leases, are unlimited in time but limited in the way that the land can be used. Few landlords farm their own property. They let it out to sharecroppers whose plots differ from year to year, denying them any security of tenure. The effect is to reduce the incentives to land maintenance so that there are relatively few trees and a great concentration on annual cropping of wheat and barley.

Some environmental effects follow from the Islamic system of taxation. Moslem rulers have raised revenue from the *zakat* tax (usually between $2\frac{1}{2}$ per cent and 10 per cent of farm income) and called on their subjects to help in the defence of the realm and in the fulfilment of religious obligations such as the construction of mosques (Levy 1974, 49). Non-Moslems paid a special and larger secular tax so that it would have been to the financial disadvantage of the state to have pressured them into conversion. The effect has been to increase the separation within the community and to drive non-Moslems into urban and professional occupations.

Although Islam restricts sexual intercourse on the Hajj pilgrimage, it has no other such restriction even during the Holy month of Ramadan, when men are enjoined to visit their wives for comfort after their daylight fasting. Overpopulation in lands with scarce resources has been restrained by the two-year period of breastfeeding during which intercourse is prohibited and all wives have equal rights to sexual intercourse in turn whatever their fecundity.

Because most Moslem lands are too dry for more than localized rainfed agriculture, they tend to have wide areas of nomadic pastoralism. The people tend to retain the character this induces when they move into areas inhabited by other faiths. This is particularly notable in the Maghreb – Morocco, Algeria, Tunisia and western Libya. A large proportion of today's inhabitants are Berbers descended from the ancient Numidian inhabitants of the area. After the defeat of Carthage in 146 BC, the Romans developed the area and built cities such as Volubilis in Morocco, Sbeitla (Sufetula) and Thysdrus (El Djemm) in Tunisia, and Leptis Magna in Libya. Agricultural development was especially advanced in Tunisia where Roman skill in the conservation and use of limited water supplies produced an agricultural prosperity based on olive cultivation that was not achieved again until modern times. The area was christianized in the early centuries but Christianity largely died out after the invasion by Moslem Arabs in the seventh century AD and suffered a final blow from the even more destructive Hilalian invasion from Egypt in the eleventh century. The people adopted Islam and a very different society followed. Steppe cultivation gave way to nomadic life. Characteristically they built their new capital not on the coast but at Kairouan 50 km inland. This reflected the Arabs' unfamiliarity with seafaring, preference for an area suitable for rainfed cereals with little land clearance, and desire for military security against the ambuscades they feared in the unfamiliar forested areas to the north.

The reversion to pastoral nomadism survived until the French occupation in the nineteenth century which brought health, security and rapid population growth. This led to revival of an agricultural system very like that of Roman times. But French rule ended in 1962. Islamic land uses came back as its faith revived. Vineyards were allowed to die. However, the return to a pastoral economy has reduced the risk of erosion because pastoralism removes less protective vegetation than does agriculture.

Over the long term, therefore, the Berber people have ultimately said 'no' to pre-Christian and Christian civilizations that have come from Europe, 'yes' to Moslem ones from Egypt and Arabia. Why is this? Two or three reasons suggest themselves. The landscape of the Maghreb more closely resembles the Near East than Europe so that invaders from the latter are immediately more at home. Also, the Arab conquerors probably brought less racial or social discrimination and had a lifestyle more similar to that of the Berbers which allowed easier assimilation.

3
The Limitations on Religion/ Environment Interactions

3.1 The incompleteness of religious observance

The main limitation on the impact of religious activity on the environment is probably the incompleteness of religious observance. This results from socio-economic pressures, political climate, available leisure and the occurrence of disease. Many people neglect dietary laws under pressure of local circumstances. The Bible, *inter alia*, condemns the consumption of pork and certain seafoods. This prohibition is observed by Orthodox Jews but neglected by practising Christians. It also condemns excessive wine drinking but Christians are probably both its greatest producers and consumers. All religions condemn theft but it is common everywhere. Pastoral people include cattle in many of their religious rituals but there is widespread cattle rustling among them for wealth and to enhance social prestige.

Many millions of people who would call themselves Christians or Moslems never attend church or mosque. Moslems should pray five times a day, but probably less than 10 per cent do this. Furthermore, the devout are often more irregular in their religious observances than they realize themselves. The rate of attendance at mass in 48 Roman Catholic dioceses in the United States was about half the self-reported rate (Chavis and Cavendish 1994, 376–81). The Church's rules prohibiting contraception and abortion are widely ignored. This reduces the birth rate and therefore the demand for houses and domestic products. As a result it increases leisure for housewives and stimulates the luxury trades.

Sometimes people neglect religious rules because they believe they are merely socially motivated developments of doctrines rather than the doctrines themselves. Buddhist rules tend to apply more to the practical running of monastic establishments than to personal morality. Hindus have special rules imposed on sub-castes beyond those which apply to all members of the faith. Moslems often marry their first cousins to avoid property leaving the family, although they recognize the dangers of inbreeding (Tanner 1964, 127–38). The Amish and other such groups inbreed to maintain the faith in spite of the health effects and the social isolation its tends to bring. It alters environmental effects by reducing their intermarriage with other groups (McKusick 1973, 1276–84).

Distance from religious authority can reduce observance of religious principles. The nearer practitioners are to each other, the greater the pressure to conformity. The further removed the recognized authority, the less important it appears. It used to be said that Irish and Breton Roman Catholics passing through the railway stations of London and Paris ceased to practise their faith. The Moslem Albanians in Kosovo would no doubt agree that inter-faith killings in Sri Lanka are wrong, but accept it as inevitable in their own long-standing hostility to Orthodox Christian Serbs. Likewise, intellectual detachment can affect commitment. Academics can be more unbiased when they are far removed from the issues they study.

3.2 Inter-religious tensions

Where there is a multiplicity of religious groups the impact of each, and also of the totality, is reduced. In the Peruvian Andes today a Roman Catholic priest competes with a number of active Protestant denominations and lay organizations in addition to the pressures of the people's ancient religious paradigms (Spier 1992, 239–65). The situation is even more complex in Western nations because the impact of each religious group is reduced by competing media and sometimes by the differences in belief among its members. Inter-religious difference can lead to tensions. These can have a profound effect on the environment by reducing mobility, perpetuating communal separations, arresting population growth and hindering economic development. Different groups begin with feelings of 'us' and 'them' towards others. This can develop into geographical separation, territorial competition and conflict. Where the numbers are approximately equal, two groups may enter into confrontation, but where they are not, the smaller group may retreat or be forced into a homogeneous area such as a ghetto.

Separations can evolve naturally and peacefully as between Buddhists and Hindus or can involve severe conflicts as seen in many countries in the past century. There has been conflict between different religions in

India, Sri Lanka, Palestine, Lebanon, Bosnia and the Sudan. In India, there has been hostility and often rioting between Hindus and Moslems since the Mogul conquest of India. Hindus sometimes invoke such popular sayings as 'if the heart is true, Mecca is in the shit house' (Parry 1994, 19). The mythical site of the birth of the Hindu god Rama is at Ayodha. The Moslems built a mosque on the site. The demolition of this mosque by mobs of Hindus led to widespread rioting.

Such conflicts tend to restrict the impact of each religion to the area of its dominance. Even after the lapse of 1000 years, the memory of the Crusades is still a factor inhibiting Christian–Moslem intercourse in the Middle East and between Latin and Orthodox Christianity in Europe. Modern Greeks still treat Tuesday as inauspicious as that was the day in AD 1453 when the Turks took Constantinople and massacred the worshippers in what is now the Santa Sophia mosque. Christian cemeteries in northern Cyprus were violated after the division of the island (Figure 3.1). Such conflicts perpetuate boundaries, dominate politics, stultify economic development and too often distort social priorities towards armaments.

An intellectual difference can be arbitrated but a difference in beliefs touches emotions which lead to entrenched positions. The more deeply felt such religious beliefs are the more the holder's political, social and economic life is involved. This makes it more likely to involve tensions

Figure 3.1 Turkish Cyprus. Destruction of Christian cemetery

which can affect the environment. This emotional component has always existed. St Bernard in his sermon in praise of the Templars wrote:

> A new sort of army has appeared. It fights a double war; the first against flesh and blood enemies; the second, the war of the spirit against Satan and vice. The soldier of Christ kills with safety; he dies with more safety still. He serves Christ when he kills, he serves himself when he is killed.
>
> (Bernard 1919, 204–6)

Recently the Prime Minister of Croatia said in a public speech that he thanked God that his wife was not a Serb or a Jewess.

There is also tension within a religious faith between those who consider themselves bound to adhere closely to its formularies and those who champion flexibility in interpretation. This tension may be augmented or muted by a sense of shame that their founder's injunctions for mutual tolerance should become so divisive.

Hinduism is so divided by its multiplicity of theistic belief systems and its endlessly subdivided hierarchy of castes, that its impact on the environment is lessened. It gives the impression of innumerable pockets of supposedly separate religious practices rather than spots which have definable and discernible edges. Some Rajputs in Gujarat see nothing strange in having both Hindu and Moslem names, which gives their social as well as their religious lives a convenient and protective fluidity.

The major tensions within Hinduism are between the castes and the Outcastes. A strike by Outcastes in Rangoon in the mid-1930s threatened to bring it very near to an environmental crisis. The more recent demand of the Outcastes for their constitutional rights has led to considerable loss of life and destruction of property. Another tension within Hinduism is between Sikhs and caste Hindus. Many Sikhs were killed after Mrs Gandhi's assassination by her Sikh bodyguards.

The split between the followers of Vishnu and Shiva that is visibly ostentatious from the parallel and upright caste marks on the foreheads of the male devout, does not seem to have led to major tensions because they themselves are endlessly subdivided into sects in which their members have their own socio-religious satisfactions.

In the past the doctrine of karma, which promised reincarnation into a higher caste as a reward for diligently fulfilling the obligations of one's religious station in life, will have discouraged protest (Levy 1974, 3). But democratic hopes and procedures have encouraged political action which has involved considerable loss of life and destruction of property.

The Jains' small numbers, nowhere in a majority, their unusual beliefs, and their subdivisions due to the complexities of religious observance over food, costume and travel restrict their influence anywhere. They do not seem to be have been persecuted even though the faith is seen as a Hindu schism.

Gautama showed wisdom in telling monasteries in their rule book (Patimokkha 1966) that they should split rather than to try to eliminate doctrinal differences by fruitless discussions, and Buddhism is tolerant of the theoretical philosophical systems into which its followers have turned Gautama's ideas. The Jewish Essenes of Qumran apparently had similar ideas about the fruitlessness of doctrinal discussions when they wrote, 'Let no member of the community answer their questions concerning any law or ordinance' (Dupont-Sommers 1961, 84).

Perhaps the practical passivity of Buddhism and to a lesser extent contemporary Hinduism, comes from the fact that they have no single sacred book. Instead they have books which are complex rather than simply directive, and are written in Pali and Sanskrit which are specialized dead languages. Also, their common belief in the transmigration of souls may make individual believers less likely to be violent over their single current existence (Levy 1974, 3).

But Buddhism has not always been so pacific in practice. In China under the Northern Wei dynasty between AD 477 and 535, there were no less than eight armed uprisings of peasants led by Buddhist monks (Yang 1961, 122). Chinese religion has also generally been peaceful. Violence has often had Christian overtones. The Boxer Rebellion of 1900 was partly anti-Christian because it was a reaction to what more traditionally minded Chinese saw as aggressive missionary activity as a by-product of European supremacy (Wilkinson 1980, 90). Monks in Burma were involved in politically motivated violence during the colonial period. This was in part because they reacted against the governing power which had controlled the support on which they depended for survival. They supported the politically motivated violence in the Saya San revolt (Cady 1958, 309–21).

The benefits conferred by Christian missionaries in the past two centuries have attracted many into the Church, especially the sick, the elderly and families with children to educate, though the results have been somewhat marred on occasion by exclusivity and a tendency to disunity. In Ngara in north-western Tanzania, the Protestant and Roman Catholic missions were demographically distinct although within easy walking distance of each other. This distinction was reflected in their voting patterns, their preference for marriage partners of the same denomination and how they coped secretly with the threats to their welfare (Tanner 1975, 161–6). The separation inevitably reduced the influence of each.

3.3 The physical constraints on religious practices

The interaction between religion and the environment is always constrained by physical limitations such as natural obstacles and by lack of resources. Mountains and seas separate people. Water availability influences patterns of religious observance and even belief. This is especially true in the Moslem world. In arid areas only pastoralism is possible without irrigation, while in semi-arid areas agriculture can only succeed where rainfall is adequate for the annual crop. In both types of environment irrigation multiplies yields but requires much more administrative control.

The semi-nomads of the Sudanese savannah are less controlled by, or even informed about, any formal religion. The nomadic Bedouin of Arabia seldom have the chance of visiting a mosque and have a free-thinking lifestyle reminiscent of the Old Testament patriarchs. At a personal level where water is abundant Moslems use it for their formal ablutions prior to praying but otherwise they use sand. Ritual cleanliness is thus related to the availability of water and perhaps warm climates.

Irrigation brings settled agriculture but requires strong administrative control over water distribution. The Egyptian farmer's approach to his faith is a response to the somewhat authoritarian agricultural pattern reinforced by the mosque and the life pattern of his village. In Bali, religious institutions may control and manage water because of its divine origins. In the Philippines and Myanmar landowners are left to make their own relations with the spirits of the fields.

The need to use buildings limits the number of worshippers and the range of possible activities. It also constrains outreach to other religions. Their construction is an economic drain on the worshipping community and uses up materials which may be in demand elsewhere (Figure 3.2). While cold climates require insulated buildings, hot climates allow large windows, wide verandas and open shaded forecourts which favour open-air ceremonies. It is significant that the Eskimo heaven is one of eternal warmth while that of the Moslem centres around lush well-watered gardens.

Religions whose rituals involve only portable materials can meet almost anywhere. They can use suburban sitting rooms, open prairies

Figure 3.2 Christian use of materials, Rouen cathedral

or even spacecraft. One may guess that some objects of religious significance travel into space with every astronaut crew. Christians and Jews in space could have prayed without being observed, but a Moslem there would have had problems in performing the necessary prayer movements and in orienting himself to Mecca.

Climate sometimes imposes seasonal or permanent migration. Seaside churches are fuller in the summer, those in skiing areas in the winter. There has always been a drift of population from colder to warmer areas as demonstrated in the USA and Europe by migrations towards the 'sunbelt'. This is echoed in church attendances. There are churches in the areas of emigration for which there is no longer any supporting population.

Some people live in areas where only certain kinds of livelihoods are possible. In the Sahara where wells are widely spaced, they form the foci of the annual religious cycle. The Sahel zone, which lies to the south of it, has more wells and enough grazing for cattle, giving them significance for wealth and security, and thus inclusion in some religious rituals.

If a community depends on cattle for both survival and social status, they will be the centre of their religion and its sacrifices, as among the Masai of Kenya and the Dinka, Nuer and Shilluk of the southern Sudan (Evans-Pritchard 1956, 221–4). These tribes accord their cattle religious status and include them in their rituals. Dinka imagination and legends 'are peopled with cattle, and they believe in sacred kine which the Nile spirit guards, only letting them pasture at night, bound to stakes, when the mist veils the banks' (Ludwig 1936, 103). The availability of meat can also control religious activity. If a community depends on cattle or pigs, they will feature in religious rituals and sacrifices. The ways in which the Christian Christmas turkey is divided up is perhaps a survival of such symbolism.

Likewise, if a community depends on crops it will tend to include them in its rituals. The Sukuma of Tanzania asperge their sacrifices with a mixture of maize flour and water, the basic ingredients for their survival in the steppe (Tanner 1958a, 225–36). This echoes the wave sheaf offerings and the Temple shewbread of the Old Testament (Leviticus 23: 10–12; Numbers 4: 7), and is represented in the use of bread and wine in Christian communion. Where trees are useful for shade or where the community depends on forest production, trees may acquire religious significance. Many Burmese villagers act on the assumption that each tree, forest, rice field and species will have its own spirit 'nat' (Spiro 1967, 46–51).

4 Perceptions of the Environment

4.1 The universality of environmental perceptions

Everyone experiences their own environment. The intensity of this experience depends to a large degree on its relevance to their livelihood and religious beliefs with which they approach it. They feel it most strongly at crisis times, such as when there are poor harvests, the deaths of cattle and children, the infertility of women or illnesses in the family. Soldiers and sailors have a widened perception of the environment when terrain or storms pose threats to their survival. This partly accounts for their common awareness of the need for religious support. In urban-industrial societies the perceived and experienced environment has less to do with survival. It can be ignored when it does not positively intrude with fog, flood and earthquakes.

In a modern city people's knowledge emphasizes their own district and its relation to the central business district, especially where there is a religious grouping as in the Jain quarter of Jaipur or the Orthodox Jewish concentration in Brooklyn. People are most neighbourhood conscious where they feel most at home, even if it is a block of apartments or a cul-de sac. It is usually in part due to religious associations. This sense of neighbourhood is likely to expand from childhood to early adulthood, then to remain static until it contracts in old age. Their experience of the whole country is more limited and infrequent.

In connection with tribal areas Semple wrote, 'The cosmography of every primitive people . . . bears the impress of their habitat.' She adds, 'The Eskimo's hell is a place of darkness, storm and intense cold; the Jew's a place of eternal fire. Buddha, born in the steaming Himalayan piedmont, fighting the lassitude induced by heat and humidity, pictured his heaven as nirvana, the cessation of all activity and individual life' (1947, 41).

Belief in the evolutionary theory, which has influenced many in all religions, can also influence environmental perceptions. It supports the view that humans may have originally graded the landscape as hunter-gatherers, giving highest value to those areas that gave a good prospect over hunting country or a refuge from enemies. A modern view of the environment is then seen as to some extent an atavistic return to this in its preference for the wildernesses of national and regional parks.

4.2 Religious awareness of the environment

Pre-colonial Amerindian hunting cultures demonstrated deep environmental awareness. Those targeting the buffalo and walrus had religious rituals which paralleled the productive process. Religious divination supplemented the tracking of game animals and hunting rituals accompanied their killing and transportation. A code of respect for animal victims guided their portage to camp, distribution, cooking and eating (Tanner 1979, 209).

There is evidence that they, like other pre-industrial peoples, had a benign relationship with their environment from which the modern world could learn. They believed that the animals on which they depended returned to be refilled in virtually infinite numbers so long as the hunters demonstrated proper respect (Evers 1949, 355–60). They foresaw that the white man would bring harmful changes. Thus they believed the disappearance of the buffalo was due to the European failure to show this necessary religious respect and not to overkilling (Krech 1999, 212–13).

When Chief Seattle of the Puget Sound tribes and his people were forced to leave their lands for an Indian reservation in 1855, he made a speech: 'All right, we'll go. But I ask one thing in return from the Great Chief in Washington, that the white man treat the animals of this land as his brothers.' He saw the environment as sacred, that breathing was an act of prayer, that what we do to other life forms becomes our own fate. He saw the white man as being in a state of profound alienation from the natural life around him through his drastic methods (Robbins 1991).

This view has wide implications. It perceptively underlines the dangers of the 'industrial' approach to the natural fauna of the world practised by all modern, especially Western, people. It compares unfavourably with the stable, ecologically sensitive approaches of the hunter-gatherer, pastoral nomad and organic farmer. Apparent scientific and economic advances may in fact be retrogressive unless, through religious perception, they accord with the natural rhythms and harmonies of nature.

The Amerindians who depended on buffalo meat and by-products had religious rituals for the success of their hunts. They understood the buffalo in a totally different way from ecologists. They believed that annually the buffalo came from under a Canadian lake so that when the herds did not return or came in reduced numbers, it was because they had not yet left their lake-bottom prairies (Krech 1999, 149). Burning the grass annually produced an excellent habitat for buffaloes, and the numbers killed, even allowing for overkills when herds were driven over cliffs, made little difference to the survival of the species. It was the European commodification of buffalo products, in which the Indians joined, which spelt the virtual extinction of the herds.

At the same time, evidence from the bison kills in the Great Plains and slaughter sites in Europe shows that these pre-industrial hunter-gatherers sometimes killed animals far in excess of their subsistence needs (Bartlett 1956). Likewise, a palaeolithic site at Solutre in France is estimated to contain the remains of over 100 000 horses (Krantz 1970, 164–70).

4.3 The influence of antecedent beliefs

All new religions adopt some of the previous cultural patterns of the people that bring them. No religious practice can be totally innovative. Theologians and prophets speak in inherited languages to express their thoughts in ways that reflect the social and physical environment of their origins. The Buddha came from a wealthy Brahman family. He is alleged to have been shocked by the sensual lifestyle of the household of the well-to-do minor ruler in which he was brought up, and he then took up the life of a wandering ascetic which was already an accepted Indian role. Jesus accepted and used the cultural patterns of first-century Judaism. Mohammed inherited the Judaeo-Christian background of seventh-century Arabia but had a widened view because he had travelled well beyond the Mecca–Medina area as an organizer of trading caravans and had been involved in over 30 violent skirmishes. The Koran shows that he was well aware of the contemporary culture and the character of the dominant political powers of his time on the edge of his Arab world, and the transitoriness of previous civilizations

Languages derive from ancient cultures and guide their users into distinct thought patterns. Thus they influence such basic processes as abstraction, inferential reasoning and categorization (Cole and Scribner 1974, 59, 122, 193). Some languages are more limited than others. Not all have an indefinite future tense and some may only differentiate between present and other tenses. Some do not allow for the translation of key phrases from others. Christian theologians stress the importance of God loving everyone, but kiSukuma has no way of saying 'I love God.' It has to be passive, 'I am loved by God', or probably, 'I am the recipient of actions by spirits.' Arabic and Hebrew have no capital letters so must use different words to distinguish proper from ordinary nouns.

Most people do not innovate much that is new in their own religious practices. They only replace them by emphasizing other ideas and practices that lie within their orbit of knowledge. The Sukuma are the largest tribe in Tanzania. Studying them over four decades showed that, although there had been immense political, economic and social changes during this period, their religious ideas have not changed much. The influence of ancestors had diminished because of population dispersal, but belief in spirits and witchcraft had increased with wider social contacts and urbanization.

Roman Catholics and Protestants in Northern Ireland avoid each other as much from deeply divergent interpretations of history as from current threats. The Battle of the Boyne dates from 1690, but its modern commemoration is seen by the Protestants as an expression of their identity and by the Catholics as provocation. The animosities of contemporary Christian Cretans against their former Moslem Turkish rulers is so heavily imprinted in their minds that they have grafted Turkish misbehaviour on to what they consider to have happened in the more recent German occupation.

Hindus retain ancient concepts of pollution and purity for religious reasons although they do not accord with modern concepts of hygiene and cause much debilitating disease. They continue to accept that dying and being cremated within the sacred boundaries of Banaras ensures liberation. With the improvement in rail and road communications the number of corpses brought there has steadily increased from a daily average of 32.4 in 1917–18 of which 9.3 per cent came from outside Banaras district, to a daily rate of 81.8 in 1989, of which 32.2 per cent came from outside it (Parry 1994, 56). They have always put partially cremated bodies into the river together with those they have immersed there without cremation. In addition they regularly purify themselves by bathing in the Ganges as often as possible, drinking its polluted water and often carrying it home in bottles (Figure 4.1). Every twelfth year millions of Hindus gather at Hardwar and Allahabad on the Ganges.

Figure 4.1 Hindus bathing in the Ganges

There can be no more effective way of spreading water-borne diseases than these practices (Rogers 1926, 59–91). They are also much concerned with ritual purity. This can be caused by the shadow of an Outcaste person, by the touch of impure materials, by food cooked by a lower-caste person, and by excreta. They are also reluctant to use communal facilities because of the ever-present danger that other users may be ritually impure in their private lives. These beliefs are derived more from inherited distinctions between the sacred and the profane than from calculations about cleanliness as such.

Arrangements for excreting and disposing of excreta are a major religious issue which can overrule the principles of simple hygiene taught by Indian public health officials and programmes for more than a century. The construction of latrines in periurban and rural areas has been advocated for years but almost all defecation takes place in bamboo groves associated with evil spirits, banana plantations, at the edges rather than the centre of fields, and along railway lines and irrigation channels (Parrack 1969, 35). Overall there is a dislike of latrines due to a combination of inherited religious ideas of pollution with a need to segregate the sexes. This segregation is probably common to all traditional societies. Sonjo women go uphill and the men downhill to excrete, while among the coastal Moslems the women go left along the beach and the men to the right.

4.4 The afterlife and attitudes to the environment

All religions have beliefs about the experience of their followers after death. If such continued life is thought to be desirable and attainable, individuals and groups will live so as to achieve it. If not, they will either change their religion or else modify the beliefs themselves.

Tribal religions often see the afterlife in terms of continuing family membership. The living and dead are a continuum with an interdependent relationship of support and influence. Since each family has its own ancestors and only periodically experiences their influence, the shrines dedicated to them are unsubstantial constructions and their influence on the environment limited and familial (Tanner 1958a, 52–62).

In Chinese religions apart from Buddhism, the plethora of local cults illustrates a general concern for the welfare of the dead. This objective and the need to control cosmic forces is thought to come through the ever popular feng-shui, which seeks to analyse how *ch'i* (cosmic breath or force) is being expressed. Changing the terrain can modify *ch'i* as, for example, in the building of an artificial hill to the north of the Forbidden City to block off evil influences from penetrating the palace. Feng-shui has long been regarded as a serious burden on the living since the disposition and conjunction of the cosmic powers can change almost hourly. Ancestral spirits can inflict misfortune on descendants who fail to share their successes by sacrificing on their graves (Addison 1925, 48).

The Chinese government commented that this concern for the ancestors was a significant economic handicap since 3 per cent of arable land in Shantung province was occupied by graves and tombs. Lumber used for coffins each year could have gone to building 200 000 houses, and burial clothing used up 24 million feet of cotton cloth (Macinnis 1973, 191).

The application of feng-shui had led to the removal of boulders and hillocks, the excavation of drainage channels, the improvement of hill slopes, the grouping of trees and pools of water, the widening of roads and the layout of winding roads and paths. In the past geomancers helped with the planning of cities as in the layout of Beijing. A change in cosmic forces once led to the removal of the city of Hanoi from one bank of the river to the other (Wheatley 1971, 419). It has also guided modern town planning. Builders must consider not only alignments but the configurations of landscape in their planning of tunnels, coal mines, roads and railway lines with their embankments. Tall chimneys and the spires of churches and mosques have all been considered bad for their neighbourhoods. The first Chinese railway from Shanghai to Wuhan was closed for a time because its poor feng-shui was thought to have resulted in the death of a peasant. The Chinese credited the British with ensuring the welfare of Hong Kong by their judicious planting of trees which blocked off bad cosmic forces (Dukes 1920, 833–5). Despite its transcendental overtones, however, the primary concern in such planning was the ordering of society in this world rather than with personal salvation in the afterlife.

Belief in *ch'i* and feng-shui has in recent years moved from macro- to microenvironments. The Chinese use it in the siting of houses and the design of gardens for the rich, the arrangement of office furniture, the height of beds, and other matters which are still within the power of individuals to arrange. It influences the siting of graves because the quiet repose of the dead is essential for the well-being of their living relations.

The Buddhist scriptures have a concept of the afterlife which tends to discourage worldly effort. They name eight hot and eight cold hells with minor hells attached to each with listed durations of 64 000 years in which each day and night is the equivalent to 6400 human years. The opposite is six sensual heavens but these have the same combination of form and formlessness, so that any knowledge of the system would require extensive study beyond the reach of almost all Buddhists (Thomas 1920, 829–33). Underlying all is the concept of transmigration. This is non-individual and soulless, a continuous fluid complex called a *natman* which changes moment by moment without becoming completely transformed. It continues for an infinite number of existences, bridging an endless number of births (Pousin 1917, 35). Its effect on the environment is most clearly seen when individuals build and decorate pagodas to earn themselves afterlife benefits.

In Christianity the scriptural teaching that death is a void until Christ returns, has often been replaced by the contrary idea that the soul survives after death to continue a permanent existence either in joy or suffering. These two views differ significantly in their implications for the environment. If death is a void, the person has no conscious existence and their body is respectfully disposed of by burning or burial and commemoration with a headstone or plaque. Such Christians do not pay much attention to the afterlife of their dead. Cemeteries, unless they are maintained by some official body, are often a study in neglect. The vast Highgate cemetery in north London (Figure 4.2) is in effect a nature reserve. Such examples, with their absence of signs of concern for the dead, fall far short of their Chinese and Japanese equivalents in Manila and Tokyo. On the other hand, belief in consciousness after

Figure 4.2 London. Highgate cemetery as nature reserve

death probably encourages a sense of responsibility in maintaining and even beautifying graves and mausoleums.

Some branches of Christianity developed over the centuries a belief in saints who can intercede with God. Such saints often have a connection with a particular town or village. This may have come through his or her association with some local activity or more rarely that they lived there or that the village possesses a relic, such as a personalized object associated with them. Most Christian villages where virtually everyone was farming, or dependent on farming, had a patron saint. If the people believe the saint can visit the earth they make the tomb a shrine and place of pilgrimage. Some saints are thought to be advocates for a particular locality whose support is assured by special forms of religious observance, such as annual festivals. These are common in Italy and Greece (Benigni 1911, 562–6). England has 936 churches and their parishes dedicated to St Peter and 397 to St Nicholas, while France has at least 3000 churches dedicated to St Martin. Many believe that these saints provide special protection to the associated villages and their inhabitants. In Croatia they have popular communal rituals for funerals and for All Souls' Day with massive displays of flowers to cover almost every grave. They grow the flowers especially for this purpose, using land otherwise available for crops (Figure 4.3). In the Middle Ages, large

Figure 4.3 Croatia. Flowers on Roman Catholic graves on All Souls' Day

cathedrals or abbeys at St Albans, Reading, Canterbury, Santiago de Compostela and Saragossa housed the remains of saints. Where the venerated person is believed to have reappeared after death, the site of this reappearance may be commemorated by a large building complex, as at Lourdes and Knock. Such locations bring pilgrims who in turn attract hostels, shops and improved communications. Knock even has its own airport.

Islam's belief in death and the afterlife reflects its Judaeo-Christian origins. It has a practical attitude to resurrection, judgement and hell. Although burial is the normal method of disposal of the dead, Moslem graveyards tend to be undistinguished. They are usually plots covered with small mounds of earth or stone, each with a crude uncarved headstone, although in wealthier areas like Istanbul the headstones are carved with a turban to distinguish them from Christian and Jewish tombs. Some Moslems believe in the ability of dead saints to visit and help humans on earth. One result is the building of memorial tombs with small square domed buildings over them. In popular parlance they are called *marabouts* in the Maghreb, *qubbas* in Egypt and the Sudan and *imams* in Iraq. Large commemorative mosques have been built over the tombs of important saints and Shi'ite ayatollahs in Kadhimain, Nejev and Kerbela in Iraq (Figure 4.4).

Figure 4.4 Moslem tombs near Cairo

The Moslem emphasis in the afterlife is on men's rather than women's sensual needs and is influenced by the sort of concepts of material comfort that would appeal to desert dwellers. The Koran refers to the rewards in heaven for believers, which include fountains, fruits of every kind, green pastures, and women lovely as rubies (Sura 55: 49; Arberry 1955). The ultimate assignment to heaven or hell is on the belief or disbelief in Mohammed's mission and on the balance between good and bad deeds. Martyrdom for the faith gives admission to heaven (Sura 47: 4–6).

The sayings of the Prophet outside those recorded in the Koran go even further, stating that paradise is made of gold and silver bricks with musk for mortar; pearls and rubies for gravel, saffron for earth. It has everything to charm the senses and delight the eye. Moslems will be given strength and vigour in paradise to have connection with many women; the power of 100 men will be given to one man (Mishkat-al-Masabih 1963, xxxiii–13). The results of these doctrines are reflected in Arab domestic architecture and civic planning. They feature fine houses with abundant water, fountains and lush vegetation with enclosed areas for women. The Alhambra at Granada with its delectable fountain courts, gardens and magnificent views to the mountains is a physical expression of these ideals.

The common Islamic belief in an instant heaven after a death defending the faith leads some to disdain the risk of martyrdom, or even to welcome it. Islamic fighters in Israel, the Caucasus and Afghanistan, and Iran's Shi'ite soldiers in the war against Iraq have shown this quality. Its practical effect has been seriously to increase the death rate among young men.

Some religious leaders have sought to allay fears of the afterlife by modifying their doctrines. This has probably had the effect of reducing funeral planning. In England, the pre-Reformation parish clergy were said to relax their teaching on moral behaviour to please their hearers (Marshall 1994). Buddhist monks are not only in sympathy with their villagers but also dependent on them for food. The realities of any anticipated Buddhist afterlife are both uncongenial and virtually impossible to achieve. This has led the Burmese and Sri Lankans in collusion with their monks to create a possibly attainable salvation and has probably increased their practical preparations for it (Spiro 1971, *passim*; Southwold 1982, 140–9).

4.5 The role of wealth

The practice of any religion is expensive in the use of time, effort and materials, all of which to some degree impact the environment. The activities involved range in scale from erecting large buildings at one extreme to minor religious activities such as crossing oneself, using the Catholic rosary or the Islamic 'worry beads' at the other.

Wide public support makes a faith wealthy. The pre-Reformation Church in Europe accumulated vast areas of landed property from tithes and pious donations which enabled it to erect magnificent churches, cathedrals and abbeys. Religious activities in Islamic countries have benefited from the ownership of waqf land and, in recent years, the immense programme of mosque building funded by the oil-rich states of the Middle East.

The Chinese have carried out much charitable work through their clan organizations. Five families in Shanghai, through their clan temples, gave to the poor, supported free medical care, provided coffins, buried the abandoned dead, bought and released animals into special sanctuaries, repaired tombs and temples, and put up memorials to chaste wives and filial sons. The Hall of Effective Care maintained a fire brigade, while the Hall of Impartial Altruism dredged waterways (Elvin 1974, 241).

By comparison, the Greeks and Romans appear to have had less concern for others in social aid and charities prior to Christian influences. The upper classes looked on their own generosity as a bargain in terms of their survival in this world and the next (Hands 1968, 15, 60–1).

Time for leisure is a vital aspect of wealth and much religious activity depends on the amount available to believers. Farmers are very busy except during seasonal periods of relative inactivity. Working fathers and women with small children can seldom spare much time. The unemployed and retired can devote more. Therefore, the wealthier and more leisured members of any faith are likely to be the main support for environmental changes, notably the construction and extension of buildings, though the resources may not come as gifts from those in their immediate vicinity. The Roman Catholics built a new cathedral in Liverpool with funds provided by the investment of past gifts rather than donations from current worshippers who would be more committed to their own parishes away from the city centre. The Anglican cathedrals in Liverpool and New York City depended on the gifts given by visitors over many years. Such giving is esteemed both by Islam (Sura 4: 114) and Christianity (Matthew 7: 2). Endowments and legacies are an important source of wealth for religious organizations, though sometimes used in ways not envisaged by the donors.

One method of church building or reconstruction is for the faithful to buy every part required by the specification piece by piece from bricks up to the heating system. The building of Guildford Cathedral was partly funded by offering each brick at a price. Some religious hospitals state that their running costs are so much per minute and invite contributions on this basis.

An accumulation of donations from small givers can change the environment just as effectively as a few large gifts. Parish records showed that some 5000 helped to build the St Bernadette shrine outside St Luke's Church in New York in 1939 (Orsi 1985, 3). Its reputation for having miraculous water creates traffic problems in the summer when among other things it is used in car radiators to prevent accidents.

Buddhist monks are supposed to beg for their daily food from the community but their monasteries have in the past accumulated landed property. Prior to the Communist revolution the entire budget of the Ching-lien Ssu temple in Beijing was met by three lay patrons in turn (Welch 1967, 211).

Rulers can be large contributors, especially when they are also a part of the religious authority. The biblical example is that of King Cyrus who not only authorized but contributed to the rebuilding of the temple in Jerusalem (Ezra 1: 1–4). Other examples are the temple complexes such as Karnak–Luxor in Egypt (Figure 4.5), Angkor Wat in Cambodia and the Potala Palace in Lhasa, Tibet. All required substantial amounts of labour. Recent examples of government support include the enormous mosques

Figure 4.5 Ruins of Temple at Karnak, Egypt

in Iraq and Morocco and the Roman Catholic cathedral in Ivory Coast. These reflect the wealth of the societies which erected them.

Wealthy people with religious motives also contribute many secular buildings for philanthropic purposes. These include hospitals, old peoples' homes and hospices. Similar motivation leads to the gift or preservation of recreational land for sports fields, parks and wider areas of coast and mountain for public access.

There are different motivations for giving. Christians would claim that they are an expression of thanksgiving. Moslems view them more as a required duty, Chinese and traditional societies as an offering to secure the benignity of ancestors. Some may retain the form but evade the reality of giving. The Chinese sometimes use imitation money in much the same way that a Nuer can substitute a cucumber for a cow, while stressing that in terms of intentions and religious value they are the same (Evans-Pritchard 1956, 203).

4.6 Linguistic and cultural limitations on environmental understanding

Language strongly influences thinking about the environment. No two languages represent exactly the same social reality. The worlds in which different societies have evolved and now live are distinct and not merely the same worlds with different labels attached to them (Sapir 1949, 160–8).

Western languages owe much to a Judaeo-Christian religious background. The Bible became the most widely used text after the first century AD, and has remained so to this day. Even when its religious teaching was questioned it still remained influential linguistically. The Enlightenment, for instance, derived at second hand from Christian thinkers, put ideas into the thought forms of European languages and stimulated the Church with such themes as the use of reason for problem solving and the improvability of the human condition (Wei-Ming 1994, 21).

Islam is an extreme example because of its ascription of linguistic as well as religious authority to the Koran. This views the environment through the language of seventh-century Arabia, although new concepts have arisen since that time. It is, for instance, difficult to find Arabic equivalents for such abstract concepts as 'beautiful' and 'interesting', with their modern English connotations, in Arabic. The Westerner usually falls back on terms such as *jamil* (pretty) or *hilu* (sweet) for the first and *muhim* (important) or *kuways* (good) for the second which have less abstract connotations in the original. In practice, Moslems supplement classical Arabic with words and concepts from other, mainly Western, languages and accept the inevitable religious and cultural baggage they bring.

In discussing the possible transfer into English of Chinese philosophical concepts, Richards wrote, 'we have here indeed what may very probably be the most complex type of event yet produced in the evolution of the cosmos' (1953, 250). The stumbling block to any understanding of Western ideas and practices is partly due to the divergences between the several thousand existing languages. They differ not only in what they can express but in the ideas behind them. There is, for instance, no word in kiSwahili, the lingua franca of East Africa, equivalent to the English word 'environment'. Of the ten words listed in a new dictionary, nine are derivatives of personal space and concerns and one refers to the world in general and its affairs.

Whorf points out that individuals see and think about their world in terms of the languages they use. Different 'thought worlds' make up the 'microcosm that each man carries about within himself, by which he measures and understands what he can of the macrocosm' and so 'we dissect nature along the lines laid down by our native language'. From his study of the Hopi language he states that there are no words, grammatical forms or idiomatic constructions referring directly to what we call time or to the vectors of time and motion as are used in English (1956, 60, 252). Steiner comments that in Hopi the observer's language shapes his perception of the processes observed although there is an overall vagueness (Steiner 1975, 91).

Some East African languages such as kiKamba do not have a tense dealing with the future in indefinite terms but convey it by the context of the sentence. This contrasts with European languages which allow long-term perceptions to become a part of thinking and therefore include consideration of the environment beyond the agricultural year.

An individual's conception of their surrounding geography is known as their 'cognitive map'. It is not an exact reproduction of reality but a schematic arrangement of all the places with which they are familiar and can place in spatial relation to each other. This cognitive map emphasizes sites which are within walking distance of the person's home. With sedentary people, including subsistence cultivators, the map is small and includes the local facilities: home, school, club, railway station, church, etc. as well as the individual's religious sites of importance: synagogue, mosque, church or temple. The map becomes increasingly sketchy beyond the weekly cycle of observances and extends to the occasional pilgrimage. Nomads often have a wide cognitive map which has little regard for international borders but would include major sites associated with their faith such as the great mosques in Arabia and Egypt for the Bedouin Arabs. Likewise, migratory workers will know places of worship often widely separated from their homes.

A religious person's cognitive map will tend to ignore locations not of the observer's own faith. The individual's estimation of any site would depend on cultural background as well as on religion. But everyone will tend to have a most vivid consciousness of the centre of their religion, such as the Sikh Temple in Amritsar, the Ka'aba in Mecca or St Peter's in Rome.

4.7 Religious art and the environment

Art has always reflected religious ideas and through this has impacted the environment. Temples, churches and mosques have always been 'books' of religious history which can be viewed rather than read. The detailed environmental history of Egypt is shown on its temple walls. The Angkor complex shows the story of the Cambodian kings and the alternations of their faith between Hinduism and Buddhism. In Christian countries, medieval and modern churches tell their religious history in carvings, statues and stained glass windows. It is possible by walking around cathedrals such as Chartres to read the story of the Bible and local church history from these features.

Oriental and Western art represent landscape in different ways, with idealized versions of human life pictured as inhabiting the environment in ways that accord with their religious world-view. Ancient Egyptian and Mesopotamian art tended to be concerned with local worship and with providing afterlife arrangements for semi-divine rulers. Indian and Chinese art often stresses domestic pleasure and felicity in stylized scenery. Oriental religious art has continued to flourish in the culture of popular Buddhism, Hinduism and non-Protestant Christianity. The West in the early Christian centuries placed much emphasis on spiritual meanings. The landscape appeared as an unimportant backdrop to biblical and classical scenes. The Reformation led to a secularization of art and a shift of patronage from religious to secular bodies in the countries which adopted it. This led to a greater emphasis on the representation of nature as of interest in its own right. It reached a zenith in the Romantic movement of the nineteenth century which introduced an idealized rhapsodic character in landscape views. It found expression in pictures such as Holman Hunt's *Light of the World*, stained glass windows and church design, a notable example being Holy Trinity, Sloane Street, London, called by John Ruskin 'the temple of the arts and crafts'. It also influenced the layout of rural cemeteries such as Forest Lawn in the USA and Brookwood in Britain. In the twentieth century, the decline in outward religious observance, the proliferation of new religious and quasi-religious movements and the advance of technology have encouraged moves into new and abstract ways of seeing and representing nature in religious buildings.

Religious statuary, ritual materials, mementoes and talismans are interpretations of religious ideas. They have a long history. The trade today produces religious pictures, images and trinkets for pilgrims at centres like Lourdes and Medjugorje. They satisfy the religious art market and influence taste, particularly at its cheaper end. Their landscape views project an idealized view of peaceful romantic hills and valleys as a home for redeemed humans. The resurrection themes in Stanley Spencer's pictures with their mules and dummy-like human figures represent one individual's interpretation of an apocalyptic environmental scene which has its own appeal.

Religious sculpture and architecture depend on the approval of its secular or religious sponsors, not so much in specifying what they want as in avoiding what they do not want. It is still unusual to depict

Jesus and the Holy Family as Asians or Buddha as beings with other than idealized and symbolic physical characteristics.

Moslems allow only religious calligraphic art and geometrical designs, but these are made decorative with many kinds of intricate workmanship including arabesques, kufic inscriptions as in the Alhambra at Granada and koranic friezes as on the Taj Mahal at Agra and large mosques. This restriction on representational art also applies to woven goods. In general Arabian carpets have purely geometrical designs, but Persian and Turkish weavers allow latitude for the inclusion of floral patterns.

Sculpture has perhaps always been the most popular form of religious art because it is durable, can be shown in the open air as well as indoors, and is readily comprehensible from every angle. When in a public place it can be seen by millions. Statues of the Buddha are not only decorated but are dressed according to the seasons, so that they are seen to be in touch with human frailties. The 30 m statue of Buddha at Bamian in Afghanistan on a main caravan route must have been seen by millions over the centuries and its battered outlines were awe-inspiring. But its message was one of inactivity.

For religious paintings to have had a comparable longevity and influence, they would have to be on protected dry stone or in a dry climate. The majority of Egyptian religious paintings were on the walls of tombs for the benefit of interred pharaohs and not for public viewing at all.

The message in sculpture can be very pointed, as in the erotic carvings on the outside of the Hindu temples at Khajuraho. One shows, for instance, a man having intercourse with one woman while stimulating four others. As well as being ostentatiously sensual and provocative, this propagates a message of population growth. By contrast, the emaciated statues of some Buddhist divines suggest the reverse. The bare-breasted voluptuous women on the natural stone of the Sigiriya mountain-top fortress and palace in Sri Lanka testify to the wealth of the culture but may not have been religious at all except for some vague connection with fertility. There are numerous temples in which the Hindu monkey god Hanuman and his monkey army are depicted as building the India–Sri Lanka causeway. This engineering feat illustrates his devotion to Rama as a model for human devotion to a god.

Illustrations on papyrus, vellum, parchment and paper were for indoor use and elite viewing. Their survival was always problematic because of natural decay and deliberate destruction. Modern popular religious poster art in Hinduism, Buddhism and Christianity illustrates an unearthly form of existence. A single Hindu poster can, for instance,

Figure 4.6 Contemporary Hindu poster with environmental themes

have a complex symbolism from which it is possible to draw a wide variety of meanings (Figure 4.6). These include veneration for the lotus, elephant and cow and perhaps erotic longing for female beauty. This religious imagery has been popularized in films such as the many episodes of the *Mahabharata*, which must have been seen by many more millions in India than view popular Western soap operas.

Religious buildings reflect certain understandings and impose them on their users. The heavens are fixed as a contrast with the ephemeral nature of human life. This generates a desire for architecture to reflect this heavenly regularity. Christian churches are cruciform in design, traditionally pointing towards the east. Mosques point towards Mecca, the direction of all prayers. Most important Asian cities, particularly the Chinese, are built on a north–south axis in line with a main processional way. The chief official buildings face south. This design is symbolic rather than visual since the vista was blocked by a succession of gates, designed more to hold in and conserve spiritual power than for defence (Wheatley 1971, 423–5). Shaker villages in the nineteenth century aimed at such a perfection of design with straight lines and exact right angles (Sprigg 1975, 95).

Overall artistic representations are foci for religious activities. They can draw in pilgrims in vast numbers as well as regular worshippers. They use up space and materials and act as the centres for the propagation of ideas which affect the environments of people who have little interest or even hostility to the particular religion involved.

5
The Effect of the Environment on Religions

5.1 General environmental effects

All environments affect religious practice and sometimes belief. As religions spread from one area to another they undergo change. Christianity was to some extent changed by passing to the many lands around the Mediterranean basin. It is different in different parts of the world. It is one thing in St Petersburg, another among the Copts of Cairo, another in Rome, another in London, and yet another in Boston. Buddhism takes on a different colour in Tibet, China, Japan and Sri Lanka. Such differentiation brings enrichment (Semple 1947, 122, 175).

There is an apparent paradox in the spread of religious ideas. Few classes of ideas bear so plainly the geographic stamp of their origin, often the circumscribed habitat of its pioneer group, and yet none have spread more widely. Such ideas travel especially far from their seed beds along established lines of communication. The monotheism of the bare grasslands of western Asia made slow but final headway against forest gods of the early Germans. The Burgundians of the fifth century received the Aryan version of Christianity from the schools of Athens and Alexandria via the Danube. They then strongly supported it against the Roman Trinitarianism which came up the Rhône valley (ibid., 27). The spread of Islam to the coast of East Africa and Zanzibar followed an Arab trade route, and its spread to South-east Asia and the East Indies was by other Arabs who followed the trade route established by the Dutch.

We see the most obvious effects in geographical locations which are in some sense exacting, notably islands, mountains and deserts. Where social organizations are small and weak, islands offer the sheltered environment which allows imported cultures not only to survive but

improve where in less protected fields they would deteriorate or disappear. When learning and Christianity had almost been wiped out by barbarian invasions between AD 450 and 800, in Ireland they grew and flourished. Irish teachers gave the Picts of Scotland and the Angles of northern England their first lessons in Christianity. They fixed their mission stations on islands: Iona and Lindisfarne (ibid., 436).

In Ireland the primitive Catholic Church underwent a distinctive development. It was closely bound up with the tribal organization of the Irish people, and lacked the system, order and magnificence of the latinized Church. But it had its peculiar tonsure for monks and its own date of celebrating Easter for nearly 300 years after the coming of St Patrick (ibid., 419).

Japan provides another example of insular effects. It has always borrowed freely from neighbouring Asiatic countries and recently from the whole world, yet everything in Japan bears the stamp of the indigenous. Buddhism was Japanized by being grafted on to the local stock of religious ideas, so that Japanese Buddhism is strongly differentiated from the continental forms of that religion. The seventeenth-century Catholicism of the Jesuits was adapted to Japanese standards of duty and ritual. But Japanese religion has never experienced a real change of heart. The core of the national faith is the indigenous Shinto cult. This has survived, wrapped in the national consciousness and lifted above competition (ibid., 414).

Mountains also have a profound influence and are a protective environment for faith. The mountaineer tends to have strong religious feelings. The hard struggle for existence makes him industrious, frugal, provident and generally honest. Statistics from mountain areas show little crime. The mountaineer carries these qualities when he comes to the plain. They are a protective environment for faith. The persecuted Waldenses, a sect who fled in AD 1178 from the neighbouring plains of Italy and France to the Alps, took refuge in the remote valleys of the Pellice, Chisone and Angrogne some 50 km south-west of Turin. There, protected from attack, they have maintained the old tenets and organization of their faith. Similarly, the mountain environment of the southern Appalachians has helped the descendants of the early immigrants to retain their biblical faith with strong orthodoxy and rigid sectarianism (ibid., 599–601). As a corollary, neighbouring mountain valleys can differ profoundly in faith. The topographic barriers reduce the possibility of mixing and encourage a particularity of belief, so that geographically close areas may differ. Neighbouring valleys in the Alps, the Caucasus and the Himalayas sometimes have different religious affiliations.

Religious dogmas tend to show a tenacity of life in mountains impossible in the plains. The Kaffirs inhabiting the high Hindu Kush mountains of Badakshan, and apparently of Pelasgic, early Greek or Persian origin, have a religion blended of paganism, Zoroastrianism and Brahmanism. One intruding faith has been unable to dislodge the previous incumbent, so the three have combined. The great historical destiny of the small, isolated, barren Judean plateau was to hold aloft the chaste religion of the desert-bred Jews from the sensuous agricultural gods of the Canaanites; to conserve and fix it, to stamp it with exclusiveness, conservatism and formalism. But when all the limitations of Judaism are acknowledged, the fact remains that the segregated mountain environment kept alive a pure and undefiled monotheistic faith (ibid., 600).

Mountains have also influenced Buddhism. Driven out of India, it survives in the Himalayan border of the country among the local Tibeto-Burman peoples and has a stronghold around the mountain city of Kandy in Sri Lanka. Also, once the official religion of Korea but disestablished three centuries ago, Buddhism has taken refuge in the Diamond mountains. Far from the main roads, a dull moribund form of the faith dozes on in the monasteries and monastic shrines of these secluded highlands. Buddhist monasticism is especially common in Tibet because its convents and lamaseries naturally recommended themselves in a country where asceticism was an obvious expedient. Families in Tibet are small, yet each devotes one or more children to monastic life.

Religion also imposes some check on the growth of mountain populations. Widespread celibacy sometimes characterizes them. The number of young Auvergnats who used to become priests was extraordinary. Many daughters became nuns. Added to emigration, this cleared the field for the oldest son to inherit. The harsh conditions of mountain life make the monastery a relatively easy option while geographic isolation nourishes the religious nature and can benumb the intellectual abilities (ibid., 582).

Emigration from poor mountain regions can alter religious balances elsewhere. Geneva, once the citadel of Calvinism, has become a mainly Catholic town due to the influx of Catholics from mountainous Savoy (ibid., 581).

Deserts also have an important effect. They tend to favour simplicity, asceticism and monotheism. Semple suggests a comparison of the deeper religion of the Buddhist nomads of the Asian plateau with the lowland Chinese:

> Up on the windswept plateaus in the boundless stretch of the grasslands and the waterless tracts of the desert, ... his ideas take on a certain gigantic simplicity; religion becomes monotheism, God becomes one, unrivalled like the sand of the desert and the grass of the steppe, stretching on and on without break or change. Chewing over and over the cud of his simple belief as on food of his unfed mind, his faith becomes fanaticism ... his big spatial ideas ... outgrow the land that bred them and bear their legitimate fruit in wide imperial conquests.
>
> (1947, 1–2)

Deserts have some connection with the origins of monotheism in general. All three great 'Abrahamic' faiths are connected in their origin and development with the arid parts of Syria and Arabia. Islamic lands now cover the steppe zone of the Old World from West Africa to western China, comprising many different ethnic groups. The monotony of the environment creates an impression of unity (ibid., 512).

But outside these extreme examples it is difficult to isolate the effect of the environment on religion from the influence of other factors. No religious practice, any more than particular social behaviour, can be due solely to environmental influences. It must always be a question of degree. Where a set of religious practices are carried out more or less continuously by a group of people within a given area over a long time, the question is easier to study. This makes such groups as the agricultural Hutterites, Seventh-Day Adventist vegetarian non-smoking teetotallers, and the Hindu Outcaste groups of particular interest.

5.2 The effect of homogeneous environments

Some cultures are tied to a specific environment in which survival techniques involve religious practice.

Dynastic Egypt was a single civilization which maintained itself on the narrow strip of cultivable land along the Nile and its delta. The annual flood and its spread of alluvial soil allowed the country to become great. The rainfall which fed the river came from faraway and unknown sources outside any possible area of Egyptian political control. It is not surprising, therefore, that it deeply influenced the religion. It was only when it gave economic stability and prosperity that the pharaohs could extend their dominion and their religion beyond their borders. Egyptian faith featured belief in many riverine animals, such as the crocodile, hippopotamus, fish, frog, ibis and other birds (Smith 1969, 313). Ikhnaton's hymn to the sun refers to both a Nile in heaven and on earth which comes from the underworld to water the fields (Pritchard 1955, 371). Also the devotional Papyrus of Ani says: 'Thou makest the generations of men to flourish through the Nile flood and thou dost cause gladness to exist in all lands, and in all cities and in all temples.'

However, despite its enormous temples, the Egyptian religion was a series of geographically separated local cults with little influence from outside and no written basis (Morenz 1994, *passim*). The book of Exodus tells us that God, through Moses, turned the river to blood to persuade Pharaoh to release his Israelite slaves, implicitly condemning its worship. This, like other plagues, struck specifically at items worshipped in Egypt, e.g. the river, sun, cattle, pharaoh's first-born (Exodus 7–10).

Other physically governed cultures had a much narrower environmental base. The Tewa of New Mexico (Ortiz 1969, 19–21) are a sedentary people confined historically and environmentally to a few villages on the tops of widely separated hills, each with its own circumscribed area of cultivable land. There is no general agreement about which of these hills is of concern to all. This diversity is reflected in their religious ideas. They classify existence into a hierarchy of categories based on the village units. The innermost structure is linked to the open spaces in the centre of each. The outer boundary of the area is four sacred flat-topped spirit-inhabited mountains. Within this are piles of stones as shrines associated with the cardinal directions.

The Ssonjo of Tanzania live in small spring-dependent villages beneath the Rift Valley escarpment. They have a monotheistic religion based on a supernatural being Khambageu living in the sun and in the local volcano, Oldonyo Lengai. He is concerned with their welfare and is worshipped at his temple in each village at festivals synchronized with their agricultural cycle (Gray 1963, 97–129). They have a similar water-related cult as the Tewa and use digging sticks which break just enough soil to plant their beans and no more. This is not an outdated backward method but the result of understanding that deeper ploughing with iron tines exposes too much of the surface soil to excessive drying. Iron plays no part of their religious concepts of survival. Water is both rationed and brought to the fields by seasonally constructed aqueducts.

The Hopi Indians of Arizona are similar. Their life depends, or recently depended, on the cultivation of a very small land area below their hilltop villages. Farming for them has always been an economic activity more than just tinged with religion. They see no conflict between the practical and religious spheres. The annual rainfall of 250–350 mm comes in seasonal flushes down a few gullies. Survival depends on its conservation in the subsoil. Even though they are now partially integrated into the surrounding Anglo culture, they do not altogether trust it. They are always aware of their tenuous dependence on their fields. Their religious rituals concentrate on the needed fall of rain (Loftin 1991, 3–11). While it appears to outsiders that these ceremonies do not, and indeed have not, caused adequate rain to fall over the years, they do not view their prayers as ineffectual. If asked why they pray, they always reply that they do so for rain, crops, health and a long life and that their prayers are efficacious. Many educated and logical people elsewhere pray for much the same reasons. There was a famous conversation between David Livingstone and a rain doctor who enquired if Europeans could be so stupid as to believe that rain could be made in this way (Livingstone 1857, 23–5).

The Hopi work the soil into deep trenches with a digging stick These trenches catch the maximum moisture, helping to develop strong root systems. They follow the methods of their ancestors who learned the techniques from the gods. The use of modern agricultural methods and iron tools would disperse the scarce moisture, lower yields and expose the soil to wind erosion. The Hopi methods make optimal use of the cultivable possibilities of their harsh environment. For them a shortage of moisture is not due to the overuse of their fields but to improper spirituality which is shown up in ignorant, greedy and irreverent interactions with the Earth Mother (Bradfield 1971). Even to kick the earth is to show a sign of poor character (Toelken 1976, 14).

Thus, both the Ssonjo and the Hopi tribes avoid using the steel plough because of the risk it poses to water conservation in the soil by exposing the subsoil to evaporation. This restriction is considered to have a primarily religious origin. Both tribes had been taught by the founder god to revere the earth as a relative.

5.3 The effect on Hinduism

Hinduism did not start at any particular date or from the influence of any particular charismatic individual. In its present form it developed in central India. The currently available documents from its early centuries are mainly abstract and probably not intended for public reading or understanding since they were written in Sanskrit, a specialized language. They give no evidence that the faith resulted from any active choice or corporate agreement on adaptation to the environment. Hindu individuals and communities thus adapt to local situations as they see fit since there has never been any central supervising organization. The people now use polluted water as much as they ever did and the consequences are reduced by civil and medical action quite independent of religion.

Vegetarianism is a Hindu, Jain and Buddhist ideal but in practice is mainly imposed when little meat is available except possibly for village feasts. It may have originated in part as a response to environmental constraints. With the dense population levels in the Gangetic plain, it makes little sense to raise cows for slaughter, when their milk is a staple, and even more, their dung is needed for fuel.

Bali is a special case illustrating the way in which the Hindu religion has adapted to its environment in ways that are distinctive and unusual. As an island it has been able to retain its traditional interpretation of Hinduism surrounded by densely populated Moslem islands near by. It is significant in having an environmentally responsive religion which clashes with scientific concepts of agriculture. The faith is closely linked to rice cultivation. It defines time as a structure composed of many interlocking cycles based on the rhythms of growth of both the agricultural and social worlds. It rejects the concept of a 'linear' agricultural cycle and a direct relationship between this and a particular god.

Au Agama is a sacred Hindu writing and a *tirtha* is a river or mountain made sacred through association with a deity or saint. In Bali, *Agama Tirtha* represents a religion of holy water. It is centred on the supreme water temple Pura Ulundanu Batur high on the one dominating mountain, and contains shrines to 147 different deities. For the Balinese, protective benevolence flows downwards from the mountain and not only does the axis of the local village run towards this point but individuals will tend to sleep in the same alignment (Grader 1969, 135).

Each farmer has a small shrine where the irrigation water first enters his fields and there are further temples where water enters a group of fields. Rituals follow the natural catchments of rivers and watersheds, ignoring political boundaries. The regional water temples set the cropping patterns and irrigation schedules. One such temple included 13 local farmers' associations divided into three groups for rotational irrigation. Each subordinate shrine or temple was associated with some particular component of the landscape. The temple rituals call into existence the task groups whose only function is to manage the rice terraces for economic production (Lansing 1991, 44–54).

In 1977 the government told the farmers to ignore the temple scheduling system and to plant as often as possible to increase yields. As a result the system of region-wide fallow periods broke down, plant diseases and pests increased rapidly, and the pesticides killed off fish and even poisoned some farmers (Lansing 1987, 326–41). In the name of science the Indonesian agriculturalists had ignored the indigenous system. Here was a clash between two parallel systems. On the one hand bureaucracy demanded a unilineal hierarchy of specific responsibilities; on the other the diffuse Balinese religious system integrated the irrigated environment with social groups through obscurely chosen temple spokesmen. The latter system proved better adapted to making optimum use of the terraced fields.

5.4 The effect on Buddhism

It is difficult to find anything in Buddhism which can be correlated to the environment or anything environmental that can be related to the Buddha. A system of thinking which denies the realities of existence and that states that any such conceptions should be progressively abolished in a succession of soulless rebirths cannot be deeply concerned with anything in the material world. In effect it advocates inaction and depersonalization. Very few people completely observe it other than the monks who have a life commitment.

Few would aim at their own total extinction, much less one which could only occur at the end of as many rebirths as there are grains of sand in the sea. It is not surprising, therefore, that Buddhists have tended, without openly denying these basic tenets, to adopt a more pragmatic approach to daily life. This adaptation is mainly personal rather than communal, and responds to local circumstances (Spiro 1971, 31–190). It appears more reserved and subdued in the colder climates of Tibet and Mongolia than in tropical Thailand, Cambodia or Vietnam.

It also influences buildings. Climate and materials impose a degree of differentiation between countries. In the cold mountain environment of Lhasa, the Potala Palace is a massive stone-built palace standing prominently on a hillside. By contrast, the Buddhist temples in Bangkok are lighter, more open and more ornate, involving much gold leaf. The climate probably affects the type of worship.

5.5 The effect on Christianity

Christianity originated in a small village environment of subsistence farmers combining cultivating with pastoralism. The lasting impression from the four Gospels is of the world of the shepherd, cultivator, artisan and tradesman who have only occasional contact with a large city. It is only after about AD 50 that the Church expanded into the Gentile cities of the eastern Mediterranean, as recorded in Acts. In its early centuries the faith spread around the Mediterranean basin and into Europe, mainly in areas depending on rainfed agriculture with stock but with relatively little pastoral nomadism. Unlike Islam, Christianity has no specific rules which would tend to link it to a particular environment, so its practice has always been less environmentally constrained. In fact, its emphasis on individual challenge and response has enabled it to adapt to all lands from the equator to the pole.

Different forms of Christianity reflect different environments. Mountain people tend to have a more austere and reserved approach. They can be slower to accept a new faith, but once they accept it, they can be more unchanging. Plains people can be more open but their greater wealth makes them more vulnerable to luxury and corruption, and probably to more openness and flexibility in doctrine. A similar contrast exists between Christian seafarers and landsmen. The former are numerous where commerce and fishing are possible and coasts indented with harbours. Because of their dangers seamen are often especially devout and independent of mainstream churches. Sects such as the Brethren are, for instance, common among the fishermen of Britain's east coast.

Different rocks impose different religious building styles. A study of church architecture in Europe is a lesson in geology. Limestone, sandstone and granite were used where available though sometimes had to be infilled with cheaper rubble. In clay lowlands builders had to fall back on brick or wood. The contrast between the limestone of the Oxford colleges with the brick of Cambridge colleges is due to the different availability of materials.

The colder climates of the north necessitate more enclosed churches and cathedrals than in the south. They also tend to have less flamboyant decoration. In their restrained and somewhat formal stained glass and statuary, the Gothic cathedrals of northern Europe contrast with the exuberant carving and ornamentation of the cathedrals of southern Europe. This is exemplified by a contrast between say, the cathedrals of Durham and Amiens on the one hand with St Mark's Venice or St Peter's Rome on the other.

Ambient climate also influences styles of worship. In northern Europe and North America religion tends to be more restrained and cerebral while in Mediterranean and tropical countries it is more colourful and exuberant. A similar contrast exists between the slower beat of church music in northern Europe and the lighter and more rhythmic African styles.

Climate and soils determine land use patterns and these in turn influence religion. Agriculture gives a more settled life than pastoralism. As a result it makes for a more settled form of religion, with permanent church buildings and a resident minister sustained by a static population. Nomadic populations often lack these. Their religion tends to be more individual and less expressed in ritual observance. This contrast is exaggerated where the agriculture depends on irrigation. Here the need for water control imposes a stricter communal organization and a more characteristically centralized and authoritarian administration. A modern example is the discipline imposed on Mormon society by the need to operate a sophisticated irrigation system.

Christians have normally interpreted the liberty of the faith as relaxing the Mosaic restrictions on diet and hygiene, but environments have also imposed controls. Countries which cannot grow wheat or vines have to use substitutes for the Holy Communion, and oils other than olive for anointing. Eskimo Christians cannot easily observe all the Mosaic dietary restrictions. Pitcairn islanders, as Seventh-Day Adventists, aim to be vegetarian but their restricted resources and island situation induce them to eat fish, although they retain the taboo on 'unclean' forms of seafood.

The outstanding example of Christian groups influenced by the environment was the early desert monasteries attractive to ascetics. In medieval Italy St Francis of Assisi preached and practised absolute identification with nature but even his immediate followers soon abandoned his approach (Dubos 1974, 122–30).

5.6 The effect on Islam

Islam was created in a physical and social environment which had two aspects: the desert and the town. Most of Arabia was almost waterless and a harsh environment in which to survive.

The desert inhabitants were mainly nomads seeking pasturage created by irregular rain. This pastoral life was uncertain so early Islam depended more on the growing towns whose wide trading networks often had little relationship to their immediate surroundings. This dichotomy of lifestyle between nomadic and urban life is reflected in the Koran and has been a continuing theme in Islam.

The landscapes of Arabia tend to favour an austere monotheism. Thirst and hunger are ever-present dangers and the traveller develops a feeling of total dependence on vast and arbitrary natural powers over which he had no control. The fatalism and the sense of divine remoteness in Islam owe something to this formative environment. 'The faith of the desert tends to be stern, simple and austere. The indulgence which Mohammed promised to his followers in Paradise was only a reflex of the deprivation under which they suffered in the scant pastures of Arabia' (Semple 1947, 514).

The habit of nomadism induced by the environment also contributed to the rapidity of its initial advance into the neighbouring and environmentally similar areas of North Africa and western Asia. Early Islam caused a great upheaval of migration. The vast empire of the caliphate, from its starting point in Arabia, spread in 80 years from the Oxus river to the Atlantic Ocean. Similarly, the rapid rise and spread of the Wahhabi clan and sect between 1745 and 1803, which resulted for a time in their domination of much of Arabia from their home in the Nejd, recalls the stormy conquests of Mohammed's followers (ibid., 512).

The cities of Arabia and neighbouring areas depended on the caravan trade between the Mediterranean and Indian oceans. Their teeming markets form a strong contrast with the intense emptiness of the surrounding deserts. The punctuation of great penury with occasional abundance helps explain the open-handed generosity of Moslems, however poor. Early Moslems prospered where there was wider political stability, low taxes and a religious code of business morality. Islam succeeded in part because it satisfied this need in providing a stable system of government administering divinely inspired laws which were easily understood and not subservient to the empires of Persia or Byzantium. It allowed a clean sweep of restrictions seen as alien or out of date.

5.7 Social and physical constraints on religious practice

Religions, whether they appear at a specific time like Christianity and Islam or evolve over a long period like Hinduism, are practised in specific environments, but there is a limit to the effects these environments can have. Social and physical constraints can strongly bias religious behaviour. But where such constraints are severe, only strong religious conviction with spiritual support can overcome them. In general severe social conditions and environments lead to a refinement of religious life.

Some constraints come from social causes. Missionaries coming to foreign countries with a Christian message have in their minds the social and natural environment from which they come. This causes a challenge both for themselves and for their converts to establish their new religious insights and practices in a new environment. For instance, traditional faiths, especially in tropical countries, are often practised in the open air, but the expansion of Christianity has frequently provided its followers with a range of climatically unsuitable buildings.

In Usukuma (Tanzania) people from each religion form groups to cultivate together even though this conveys no particular economic advantage. Hospitals and schools run by particular Christian denominations are an attraction to the indigenous people. However, a shortage of Roman Catholic priests, nuns and brothers has reduced the possible environmental influences of that faith. But Seventh-Day Adventists in the same area with their moral and social cohesiveness and without the temptations of drink and dancing, have been a positive influence on the growth of cotton cooperatives. In turn the replacement of food crops by cotton growing for cash was a major environmental change. It led to the importation of food into what had previously been a large farming area feeding its people and selling their surplus.

Moving to a new area is easier when the community religion is money-based rather than environment dependent, or when its practices do not require a specific lifestyle. Hindus, Moslems and Sikhs in Britain can find employment and commercial opportunities without too much strain to their religious practices and needs. They create their own temples and mosques through which socio-religious opportunities and needs can be created independently of the urban environment in which they live. They are scarcely noticeable environmentally.

Some constraints are environmental. A specific religious practice may be tied to a range of places in which it can be practised. Although Moslems can pray anywhere, the obligation to go on the Hajj to Mecca once in a lifetime has a specific environmental requirement. Devout Moslems overcome this by travelling from all parts of the world. Some peoples are so constrained by the linkage of their religion with the environment that they could hardly live elsewhere. An example are the Mbuti pygmies of the Congo whose whole life depends on the forests in which they live and which is the home of their benevolent forest deity.

Too close environmental controls inhibit the expansion of a religion. This is well illustrated by the complex polytheism and rituals of Balinese Hinduism closely integrated into their terraced and irrigated rice growing. It is not suited to other areas. The Amish interpretations of the Bible seem to have brought rigidity to their social and agricultural system, constraining the possibilities of being able to expand and to retain the self-supporting structure of their communities. Such expansion has started to break the close and successful relationship between their religious practices and the environment to which they had become acculturated.

Norse settlers had to leave Greenland because they could not adequately adapt to the environment, much less copy the lifestyle of the Eskimos who survived there. Their models for churches came from the well-forested environment of Scandinavia and could not slot into a treeless one with very limited fodder for keeping cattle over a long winter (McGovern 1980, 245–75). Their stone churches with wooden roofs in a treeless environment are now a mute testimony to the difficulty of transferring older practices to a harsher new environment.

5.8 The effects of modern industrial and post-industrial environments on religion

In 1850, Western society was still mainly rural and the largest cities were in Europe and North America. People seldom left their native areas throughout their lives. Today, much of the world is urbanized and the biggest cities are in Asia and Latin America. Economies are supported by relatively decreasing agricultural and mining sectors, a growing industrial sector and a rapidly increasing service sector.

The most obvious religious change is the decline in the outward observance of religion. This is a worldwide phenomenon and occurs even in subsistence economies because of increasing leisure and the questioning of traditional ideas though widely accessible media. But the picture is not simple. Regular church attendances in Europe have fallen from perhaps a near majority in the past century to less than 10 per cent today but they remain high in North America. In both these areas there has been an 'individualization' of religious practices and a proliferation of alternative religions. There has been a spectacular growth of small church groups with clear-cut beliefs and an absence of moral equivocations. The movement of populations has brought many Asians to the West who have brought their religions. The ease of travel has led to an increase in pilgrimages and in the understanding of other faiths. The environmental effects are mainly seen in the increased movement, the changes in the use of religious buildings, and the changing social customs and dietary habits.

5.9 The effects of population increases on religion

Population increases in the past century have been worldwide but unevenly spread. They are largest in the Third World: Asia, Africa and Latin

America. The growth in Asia has not been accompanied by any great change in the relative proportions of the main religions with the exception of Hindu Outcastes moving into Buddhism in large numbers. Africa has seen increasing Christianization mainly through independent churches, which has correlated with a 'Westernization' of lifestyles and a move towards the consumer society. This has to some extent evoked a revival of Islamic self-consciousness and a spread of the faith around its traditional areas as well as through outward migration to Western countries.

These changes have had environmental effects in spreading Christian and Moslem behaviour patterns beyond their historical core areas. The movement of Jews to Israel from both Europe and Arab countries to form a state with a population of over 5 million has imported into that country lifestyles which originated elsewhere and reflect the socio-economic practices of many source countries. It was mainly immigrants from Europe, for instance, who conceived and operated the kibbutz programme.

Such movements usually involve urbanization and religious ghettoization as between Moslems and Hindus in India, Hindus and Buddhists in Sri Lanka, and between Protestants and Roman Catholics in Northern Ireland. Many Moslem Fellata tribesmen coming from Nigeria on the Hajj have been unable to afford the whole journey and have settled in the Sudan en route. There they have formed their own villages separate from those of the local population.

This tendency is to concentrate the lifestyles and land uses of each group into discrete areas. This has inevitably caused local tensions between the immigrants and the receiving countries. It has occurred with North African Moslems in France and Spain, Turks in Germany and Sweden, and south Asian Moslems in Britain. In extreme cases population pressure and political fears have led to prolonged conflicts between religious groups, as in Israel and the Balkans.

6
The Impact of Religious Behaviour on the Environment

6.1 Enthusiasm and drive

The dynamism induced by some religious beliefs has an important influence on human uses of the environment. Christianity is the outstanding example. Christian missionaries have always accompanied and strengthened trade. Jesuits accompanied fur traders in the wilds of central Canada; Spanish priests accompanied the gold seekers to Mexico and Peru. The United States owes Hawaii largely to the efforts of American missionaries. Thirty years after their arrival in the islands, they had secured tracts of land and held all the important offices under the government. This had far-reaching consequences since it laid the foundation in the early twentieth century of the landed aristocracy of planters. Their sons and grandsons took the lead in the revolution of 1893 and in the movement for the annexation of the islands to the United States (Semple 1947, 98–9). Christian missionaries in those parts of Africa colonized by Britain had an educational influence which played an important role in the adhesion of these territories to the British Commonwealth after their independence.

6.2 The use of time

All religious activities use up time and involve choices both for the person and the group. The time spent on them relates to the importance accorded the activity. Visiting the mosque, temple, synagogue, church or going on a pilgrimage means forgoing other activities. Even home devotions have this effect. Some traditional African ceremonies in the dry non-cultivating season may take all day in fits and starts, although they only take time from agricultural work in emergencies. Traditional cultures may have concepts of time which vary with the necessities of life. The Nuer of the Sudan consider that time is slow in the dry season when there is not much to do and quick in the rains when life is much more busy (Evans-Pritchard 1939, 189–216). Religions vary in the emphasis they place on time conservation. Viewing it as a commodity in short supply requiring care and conservation tends to be peculiarly Judaeo-Christian, and this has led to the shortening of services.

Heavenly bodies define the year, month and day. Their observation and measurement have always been a religious concern. Ancient observatories noted the movements of the sun which controlled seasons and of the moon which controlled eclipses and tides. In all religions festivals marked the seasonal cycle of seed time, harvest and fallow. The month is the basis of business accounts and in Islam it is the lunar month rather than the year which is the basis of long-term calculations. It is annually emphasized by the Ramadan fast. The day is the framework of all religious and secular activity. The week is the exception. There is nothing whatever in nature to record it. It solely affects humans and is quite universally observed. It is punctuated by a regular rest day, likewise more or less universally followed, or at least acknowledged. Most Christian denominations reserve Sunday, just as the Jews and some other Christians reserve Saturday as a period when very little should be done outside their devotions. Moslems visit the mosque on Fridays. This weekly rest day has environmental significance in interrupting business activity by relative quiescence. Working, eating, travel, as well as religious activities, are different on that day.

The day of rest may involve sacrifices. The Christian Eskimos of the Bering Strait were prohibited from hunting on Sundays. Three of the best days for hunting walrus happen to be on three Sundays in June so that overall the non-Christians appeared to have some advantage in getting more meat into store for the coming winter (Ray 1992, 221).

It is mainly monotheistic religions that run by the clock. Churches have sundials and clock towers and mosques often have large wall clocks to synchronize the times of prayer. The main mosque in Xian, north China, for instance, has a board with five painted clocks on it to give the times of required prayers and a live clock to give the actual time. In Saudi Arabia many businesses arrange and advertise their working hours around these prayer times.

In any country there are many individuals whose main use of time is for religious practices. This includes Christian priests, ministers, monks, nuns and comparable dedicated staff in other faiths. The total in large countries such as India, Thailand and the United States runs into hundreds of thousands, if not millions.

Monasteries slot hours of prayer into farming procedures. Early Cistercians were so successful that the monks became independent of outside support and were able to devote their non-working hours to individual and collective prayer. Thus a substantial use of time for prayer did not prevent the development and continuation of successful farming.

Others committed to the religious life do not belong to closed institutions. Priests, pastors and rabbis spend much time on non-religious activities. A work study of how Roman Catholic priests in Tanzania spent their time showed that their religious activities decreased after Vatican II. On the other hand, many employed in secular capacities in religious institutions use time in ways that can be defined as religious. This includes secular administrators of religious foundations and hospitals, and a whole web of subordinate cleaners, cooks and clerks.

Of the Five Pillars of Islam, the Hajj is the one with the most significant environmental impact. The Ramadan fast reduces working efficiency especially when the month falls in a hot and thirsty time of year or when it coincides with long summer daylight in high latitudes. It alters the times, but not generally the total amount, of food consumed. Almsgiving, the affirmation of faith and five daily prayers can be carried out anywhere. The Friday prayers at the mosque encourage but do not demand the building of a mosque as a locus for the communal expression of faith. Prior to the use of air transport the Hajj often took several months by steamer and even longer overland. Those who came from West Africa on foot via the Sudan took more than a year as they had to earn their passage stage by stage.

Christianity has always had movements of the same type. In the Crusades many men were away for years. Lourdes receives more than a million pilgrims per year each requiring about a week for travel and stay. This localizes travel routes, accommodation and food consumption.

6.3 The use of space

Religious practices and institutions use large amounts of space. This is seldom intensive as much of the ground area remains semi-permanently unoccupied. Churches, mosques and temples are in the middle of human settlements and include supporting facilities such as housing, utility halls and cemeteries. They sometimes seek high ground as indicated by ancient hilltop shrines and ziggurats. In Ethiopia, as in southern Europe, churches are usually conspicuously on hilltops.

Planners of religious precincts calculate their requirements from maximum anticipated use, allowing for extreme needs at exceptional times. A Christian church is often relatively empty for its main Sunday service but will have few vacant seats in the main services at Easter and Christmas, a fashionable wedding, or the funeral of a popular figure. A cathedral with its side chapels, altars and accumulation of monuments with only a small area occupied by seating is often only used for religious purposes a few hours a week. The central mosque of Islam surrounding the Ka'aba in Mecca has a circulating mass of thousands of worshippers during the Hajj but many fewer in the non-pilgrimage months.

Burial grounds were originally subdivided into sites for single graves but most now have layers of burials and contain many tombstones, mausoleums and other memorials. Some become deserted and unknown until rediscovered by archaeologists. Some are developed as parks or playgrounds with surviving memorials stacked along the walls. Before tombstones began to be made in permanent materials, the recycling and associated renewing of the sacredness of the land was a continuing process with new burials and their associated rituals.

Where cultivable land is limited, graves have no permanent memorials. In Crete, for instance, ossuaries hold the bones of the village dead. Increasing wealth and the return of emigrant Greeks to die in their homeland has led to the marking of graves with marble headstones. This has caused cemeteries to encroach on to needed agricultural land.

Another factor affecting the use of burial space is the attitude of worshippers to their departed relatives. In the Western world this is restricted to personal memory. Ancestor propitiating societies have a vital interest in sustaining an active connection with the dead. In Manila the Roman Catholic and Chinese cemeteries are adjacent. The former shows every sign of neglect after the ostentation of the funeral. The latter shows active and continuing preoccupation of the living with the dead. While the Japanese, Taiwan and overseas Chinese constantly not only remember their dead but involve them in their current affairs, most Christians end the memorialization of their dead quite quickly except for some rituals on All Souls' Day.

This Chinese ancestor worship has brought an ever-present and powerful element of intimate family coherence, uniting it with the environment within an all-pervading I–thou relationship (Gutkind 1956, 1–44). In the province of Kansu, graves cover many square miles of the cultivable plain. This is because the ancestor cult and the general reverence for ancestral graves were stronger even than the pressing need for the careful preservation of the soil, in spite of its insufficiency for most Chinese peasants.

Religious cemeteries may unintentionally become nature reserves. There are 103 cemeteries in London of which that in Tower Hamlets, covering 33 acres in Bow, has been described as astonishingly overgrown (Meller 1999, 280). Part of Kensal Rise cemetery is fenced off as a scientifically managed nature reserve. Highgate cemetery in London has reverted to the type of woodland which it may have had under the Anglo-Saxon swidden cultivation cycle. During their active life these cemeteries contributed substantially to a horticultural, rather than a natural, environment. 'Go behind the scenes at Highgate. You are in a maze of beds and glass-houses, working in and about which is a regular staff of twenty-eight gardeners. Merely for bedding out some 250,000 to 300,000 plants are raised every year' (Sims 1906, 82). Churchyards make a significant contribution to variations in the environment. It is estimated that the average English churchyard contains the phosphate from the bones of 10 000 burials (Rackham 1986, 344) and is crowded with wildlife.

Some religions dedicate areas to disembodied and undefined spirit forces. This includes the sacred groves common in many cultures surrounding springs or along watersheds where people may not cut anything green and only collect dead wood. The large numbers of sacred groves in India preserve plant species which have become very rare or extinct elsewhere (Gadgil and Vartak 1975, 314–20). Piles of stones, constantly augmented by passing travellers, often mark mountain passes and promontories.

The Pare of northern Tanzania live in intensively cultivated mountains where little forest remains (Kimambo and Omari 1972, 111–21). Their sacred groves vary in size with the importance of what they symbolize and the size of the community. Shrines marked out by a sacred tree have an almost totemic value for particular clans. The sites of initiations and rain-making forests often cover large areas of land. With the development of state rituals, such places become social rallying points and useful foci for political centralization. A spring near to Bukoba in Tanzania, which Roman Catholics now regard as having sacred Christian properties, is the site of an earlier traditional sacred grove.

Hindus consider the Ganges to be sacred from source to mouth. Certain areas along it are especially sacred and are used for the immersion of dead bodies and ritual washing. This sacredness does not prevent the wholesale pollution of the water by human waste nor has it ever motivated believers to preserve its banks from erosion.

In the Foggia plain in south-east Italy there is an area of oak and scrub around an ancient shrine to the Black Madonna commemorating the believed appearance of the Virgin Mary in an oak tree. It led to the building of this isolated woodland church (Bradford 1957, 89). The spirits living in these sacred groves are sometimes thought to protect water sources. When the people lose this belief and cut down the protecting trees for firewood, they reduce the water supply. In Loksawm village in the Chin region of northern Myanmar a well completely dried up and the people had to move (Stevenson 1943, 45). This parallels the difficulties in Peru where secular water policy divided villages on the bases of landownership. This conflicted with the local belief that supernatural forces, represented by the images of Catholic saints, adjusted the flow as a result of believers' religious rituals (Holmberg 1952, 113–23).

An even more complex relationship with the land is when the soil is seen as holy in itself and thus is wounded by the blows of agricultural implements. This makes some pastoral peoples such as the Masai in East Africa reluctant to undertake cultivation or contribute to road building.

6.4 The use of materials

There is a distinction between organic and inorganic materials. The former include animal and plant products. Blood sacrifices form no part of contemporary Western religion, but most traditional religious systems included them periodically. The early Mexican temples were described as black with the crusted blood of recent human sacrifices. The Hajj involves the transportation to Mecca and killing there of not less than 100 000 animals in a short space of time.

The sacrifices of traditional religious systems may have a culling effect on overstocking as well as adding protein to the local diet. This change of diet is comparable to the feasting on saints' days and Christmas which has been constantly a feature of Christian societies, just as celebrations for the Prophet Mohammed's birthday and the end of Ramadan always feature in the lives of Moslems.

Vegetable products have always formed a part of Christian life in harvest festivals and in tithing for the maintenance of their clergy. Mormons as a religious obligation keep six-monthly reserves of food in their houses for emergencies in part as a memorial to their persecuted and frontier past.

If it requires 50 acres of oak trees to build a wooden warship, it is not difficult to imagine the enormous quantities of high-quality timber that must have gone into the building of medieval churches and their interior decoration, rood screens and seating, as well as the webs of scaffolding necessary for their construction. The use of wood for cooking for pilgrims must have substantially reduced woods near to their routes and destinations.

The amount of organic materials expended on the dead at the indirect or direct expense of the environment must be enormous. It varies from the massive arrangements for the future lives of pharoahs and Chinese emperors to the poorest modern family which wants to spend money and materials on giving their loved ones a good send-off and even taking out insurance for this purpose. The country around Banaras is denuded of trees and domestic cooking relies on dried dung. Contractors must bring wood for cremations from more than 60 miles (96 km) away. Only Cistercians bury dead monks in their everyday clothes without coffins. These contrast with the Parsi practice of exposing their dead for consumption by vultures.

In Thailand, Myanmar and Sri Lanka, religious outfitters provide the orange robes, fans and other monastic apparatus to the large number of young men who enter the monastic life for a period. This is especially significant because the giving of such robes at public feasts marks the most important occasion of family life. Otherwise monks are inconspicuous except in officiating at funerals. If overall a million young men go through this monastic rite of passage and each receives two sets of cloth, this represents a substantial use of organic materials. There are similar outfitters in Jerusalem and Dublin who advertise in religious journals for clothing associated with religious vocations.

No religious institution is without specifically religious literature available for sale or free distribution. The Bible and the Koran are easily available. The former is produced in many hundreds of languages at subsidized prices. The full Bible in Swahili is sold at a price which would not buy much more than a small paperback book in the same language. The production of devotional books, pictures, prayer cards and promotional literature is now a vast and growing industry involving a large consumption of wood pulp.

The main inorganic materials for religious purposes are stone and brick for buildings. The former consumption of stone was enormous in temples and cathedrals. The feat of transporting the granite obelisks from Aswan to Luxor was remarkable whether done by river or land. The transport costs of religious construction work have changed over the centuries as machines have replaced human muscle. Pyramids, cemeteries, temples, monasteries, memorials and cathedrals are the surviving evidence. Their construction has continued with little let-up to the present day. Recent examples from come both developing and developed nations.

Such religion-inspired constructions reflect economic well-being. English ecclesiastical history illustrates this. The churches in medieval East Anglia and in the large cities of Victorian England followed prosperity. In Madras waves of major Hindu temple construction occurred in the periods 1640–1800 and 1960 onwards, both being times of rapid economic growth (Waghorne 1999, 648–96). This provided environmental 'breathing spaces' in an expanding city, accommodating both the economics of contemporary progress and the needs of a small community social system (Wallerstein 1991, 139–46).

The vast new Roman Catholic cathedral in the Ivory Coast used imported materials. There is a new mosque in Marrakesh, and new cathedrals in Liverpool and New York, as well as the reconstruction of old ones demolished by bombing in the Second World War. There are also new Hindu temples in Britain and North America resulting from the wealth of the communities. Oil profits have gone to construct enormous mosques in in Baghdad, Saudi Arabia and London and to subsidize their construction elsewhere. These are in addition to the construction of hospitals, universities and other educational institutions in Moslem countries which benefit their whole populations, whatever their religion.

These large mosques, churches and temples use up considerable areas of high-value land in all cities. Any aerial view shows the acreages they occupy. There are as many Christian churches in a given area of agricultural England as there are Buddhist pagodas and ancillary monastic buildings in a comparable area of Myanmar. In fact the occupation of space in Christian areas is probably greater than in comparable Buddhist or Hindu areas because burial is preferred to cremation. While Communist governments for a time reduced the number in use by demolition or conversion into museums, political change has restored them to religious use. In China there has been a surge of new religious constructions. Even during the height of the Cultural Revolution the main mosque in Xian and the Roman Catholic cathedral in Beijing remained intact.

Churches, mosques and temples are complex buildings requiring many different materials. This is related to their characteristic architecture. The main body of the buildings is usually a high hall but a specific characteristic is the use of towers, spires, minarets and domes.

Towers became important in the Romanesque and Gothic periods in Europe and have remained characteristic of church architecture since. They are usually built of stone and require buttressing. They are for observation, to hold bells and to show their purpose from a distance. They are often crowned by spires, also made of stone, which make a spectacular visual culmination of the tower and symbolize the heavenly aspirations of the worshippers.

Moslem minarets are likewise built of stone where it is available. They vary from thick, squat spiral ramps such as the brick construction at Samarra (Iraq) to soaring, delicate, pencil-like spires. There can be anything from one to six per mosque. They serve as landmarks of Islam, visible from afar and giving the landscape an Islamic character.

Buddhist pagodas are multi-storey stone structures usually associated with temple complexes. The design originates from the stupa of ancient India, erected over the relics of a holy man or king. They are often bottle-shaped. Designed as monuments, they have little usable interior space.

Funerals often have involved constructions. They have at times required an extravagant use of materials. One example is the buried terracotta army near Xian in northern China, east of the supposed tomb of the Emperor Qui-Shi-Huang, which were constructed 221–210 BC in order to guard him in his afterlife. It consists of over 8000 warriors and 40 wooden chariots. The creation of such an enormous number of life-size figures of men and horses would have used not only clay, the least important ingredient, but also pottery moulds, enormous quantities of wood, and the food and time of a large number of skilled and unskilled workmen. This is noticeable because of the massive and unusual scale of what was done for this one emperor. But the burial of materials to benefit the dead occurs on a smaller scale elsewhere as in Egyptian tombs and the famous death pit at Ur.

There is a worldwide industry to produce memorials to mark graves. These are usually carved in stone. The Chinese devotion to ancestors has led to planning the preservation of graves in fields and cemeteries. In Western cities even with the reuse of burial space, cemeteries represent a large use of stone. Even Tokyo, with its practice of cremation, has areas devoted to memorials (Figure 6.1).

Candles burnt at Christian and Buddhist shrines use up materials. The heat is not utilized except indirectly; but the sites are a mass of flickering light all the time. Many have continued such regular candle burning for many years. In the long term their fumes blacken church interiors, involving costly cleaning. Hindu and Buddhist funerals burn huge paper constructions and have to make special arrangements for them to

Figure 6.1 Memorials in cemetery near Tokyo, Japan

pass under power lines on their processional way to cremation (Figure 6.2).

Religions have always used materials for items of piety. The first mould for the production of religious artifacts to sell to Christian pilgrims dates from about AD 300. The Koreans export images of the Virgin Mary to the West. In the United States alone the value of religious artefacts produced annually for sale is several billion dollars. In the late 1970s an entire Ladino neighbourhood in San Cristobel, Nicaragua, specialized in the production and sale of ritual goods (Smith 1977, 14). The road from the airport in Bangkok passes factories producing concrete spirit houses and large 4-foot metal images of the Lord Buddha. These can be seen outside offices, police stations, factories, hotels, on the verandas of apartment blocks and in fields, gardens and mountain clearings. Devotees constantly replace the offerings of flowers, incense and fruit in front of them. The contemporary decoration of such statues with gold leaf is so thick that it rounds off the features. This continues the historical practice of decorating religious images with jewels.

The Taiwanese government has recognized the disadvantages of such religious expenditure in time and materials in an island economy dedicated to modernization, and has discouraged excessive festivals by ordering them to be carried out on similar days. Taiwanese have, however, largely ignored this ruling, arguing that such festivals contribute

Figure 6.2 Cremation container for body, Bali, Indonesia

to their sense of well-being and thus help expand the economy (Sangren 1987, 197–9).

6.5 The use of transport

Religious construction involves moving materials and workers. This was much harder in the past without paved roads and in the absence of mechanical aids. The Forbidden City, home of the Chinese Emperor and Son of Heaven, contains a 200-ton slab of carved marble brought from a quarry some 80 km away on a sledge This could only be moved by flooding parts of the track whose freezing made a firm surface. Most of the innumerable blocks used in the construction of the Pyramids came from quarries 16 km from the site, but some came from the Aswan granite quarries over 600 km upstream. An even more remarkable move was that of the Olmecs of Mexico who transported multi-ton blocks of granite over 160 km to their present sites on the coast and sculpted them with no knowledge of metal.

It is not known how ancient peoples moved these great weights. They constructed the Luxor obelisks in Aswan 200 km to the south, and then either floated or dragged them over this distance. There are over 2000 stone circles and rows in Brittany and the British Isles which seem to have had a religious purpose. The best preserved and most widely known is Stonehenge. It is built of large blocks of sarsen (an iron-cemented sandstone) from nearby downland but also includes many boulders of crystalline 'bluestone' from the Prescelly mountains of Pembrokeshire. The transportation of all would have demanded vast effort. It is believed that the bluestones were probably transported by sea from Pembrokeshire to the Dorset coast and then brought up the Salisbury Avon and thence overland to the site. Some of the vertical sarsens are bridged by horizontal pieces of vast size which have been morticed into their positions. This would have demanded lifting and carving as well as transportation. But the gradients in England are mild in comparison to the precipitous inclines which had to be circumvented with heavy blocks of stone for the creation of Inca temples in Peru. Meso-American pyramids are made of bricks and not of stone. This involved very large numbers of Indians to make and carry the bricks.

The more important the religious centre or focus, the bigger the importation of materials required. An insignificant spring at Lourdes has turned into a primary complex of special buildings and an even greater complex of secondary ones to house and service more than a million pilgrims each year.

Some religious sites were artificial. Cities such as Babylon and Ur had vast brick ziggurats which seem to have been aimed not to be tombs but to provide a site for religious services at their summits. The largest artificial hill in Europe is Silbury Hill, near Avebury in England. It probably had religious significance although its purpose is unknown. It was built by heaping up local materials to a height of about 40 m. The wonder is not only the transportation technology but the complex organization developed to achieve it.

A vast artificial pyramidal hill near Xian in northern China, presumed to contain the tomb of the Emperor Qui Shi Huang, was made by his son Qui Er Shi. It is reputed to be the largest single human construction on earth. It was originally 115 m high with perimeter walls 6210 m long. It took 38 years to construct with a labour force of possibly 700 000 carrying earth in baskets (Sima 35–45 BC). It stands out in a flat heavily cultivated plain as an obvious reserve for plants, birds and animals which might not survive elsewhere. It completely dwarfs the Pyramid of Cheops, let alone Silbury Hill.

The second aspect of transportation is the movement of people to religious centres. Time and motion studies show that pilgrimages take up a vast amount of material and human resources. Huge numbers travel to Banaras, Mecca, Jerusalem, Rome, Lourdes, Mexico City and the four sacred mountains of China every year. They require food, housing and transport. It seems likely that they employ at least an equal number of non-pilgrims in service industries to provide for them.

The most important must be the Hajj since so many pilgrims are concentrated in the one small area for a short time. There are special travel agencies both for this and for Christian pilgrims which have existed since medieval times. In the period from 1875 to 1910 the annual pilgrim caravan from Damascus to Mecca had a regular demand for between 4000 and 20 000 camels (Grant 1937, 229). In the middle of the twentieth century steamships replaced dhows carrying pilgrims from Suakin to Jiddah. This enabled larger numbers to come and increased the distance they could travel without serious difficulties. It probably reduced the cost and effort required for prearrangements. Aircraft have further transformed the situation. Many specially chartered flights bring pilgrims to Saudi Arabia from as far afield as Manchester, Calcutta, Ibadan and Djarkarta.

Transporting the millions of pilgrims within the sacred area of Mecca demands extensive resources. In 1974 four transport companies alone were providing 2550 cars and 2957 buses with another 2000 ordered (Bushnak 1978, 87–115). Traffic jams have sometimes taken days to clear. Road accidents have now replaced infectious diseases as the main cause of Hajj deaths. There are also large numbers of extra police, immigration officials, medical staff, drivers, lodging house staff, food preparers, guides and shopkeepers apart from the extra aircraft. These are in addition to the equally large number of providers of these services en route and in the countries of origin.

There are parallels in India. The creation of a railway network has enabled ever larger crowds of pilgrims, now estimated at several million, to attend the Kumbh Mela riverside festival every 12 years at the junction of the Ganges and Jumna rivers. In 2001 estimates of attendance ranged up to 70 million. In 1977 up to 3.5 million of these pilgrims visited Banaras as well. In Uttar Pradesh it is estimated that there are 553 annual Hindu festivals each attracting more than 10 000 people (Banerjea 1951, 17–40).

In Sri Lanka during the Buddhist pilgrimage season large numbers of specially decorated buses take pilgrims to the main centre in the south of the island. There are numerous examples of similar transport concentrations moving people to Santiago de Compostela, Medjugorje, Fatima and Knock as well as in Latin America. The car parks of any church in the Western world shows that the use of transport even on a local scale is now a necessary adjunct to worship. There are also special means of transport for particular types of pilgrims. Ambulance buses ply to Lourdes and in 1985 special vehicles transported 12 000 corpses to Banaras (Parry 1994, 57).

The enormous overall increase in pilgrimage going is, however, not just the result of the availability, cheapness and ease of transport. It is also due to growth in the number of sites. Even in Israel, with its sites of national worship in Jerusalem, immigrants particularly from North Africa have created new pilgrimage centres for rabbis who have died in the 1980s. They have also set aside mini-sites such as a single room dedicated to a saint buried in the Atlas mountains which attracted 15 000–20 000 on his anniversary (Sharot 1996, 120). The increase in this type of individual spare-time religious involvement increases transport, food and lodging requirements.

6.6 Domestic religious practices and the environment

The home, nuclear or non-nuclear, is both the centre of current religious activities in their most persistent and regular form and the locus for training the next generation. It is from the family home that people normally start, continue and terminate their religious activities.

Religious practices in the home start in early childhood. Where the homes of a particular religion are concentrated in one area, they have a cumulative effect on the environment. Preconditioning into attitudes to people of other religions is part of growing up. But it has led to hostility between Roman Catholics or Protestants as in Northern Ireland and between Palestinian and Jewish children in Israel. This preconditioning becomes stronger when children go to a denominational school even if it is only shown in inter-school competitive games or debates. The overall environmental effect is to deepen the segregation of the communities.

Religious shrines modify the domestic environment. Many houses in Japan, Thailand and Myanmar have shrines and the homes of Roman Catholics and Hindus everywhere have popular religious pictures and calendar art as an encouragement but also as a constant reminder of the need for commitment. Some homes in Madeira augment these pictures with large photographs or engravings of long-dead local clerics. The mud walls of a farmer's house in Luxor had photographs of Egyptian Moslem leaders and pop stars.

Many African tribal and Chinese patriarchal family compounds contain shrines to their ancestors (Figure 6.3) and view their land as family rather than personal property. Early religious instruction in the home also has long-term impacts on the environment. Children learn to observe the religious holidays of their elders. Young Jews develop an understanding of laws about the Sabbath, hygiene and the relation of their people to the land of Israel. Moslem children grow up with the knowledge of how inheritance is fixed proportionately. For this reason they are encouraged to look on their first cousins as potential spouses so as to merge inheritances.

In Myanmar devotees give daily offerings and seasonal changes of clothing in shrines to the Buddha, and perhaps also in rural settings to the local spirits. This is the local equivalent of those given to the enormous Buddha statues at Ayuthya in Thailand. The use of incense is also commonplace in Eastern temples and homes and is popular in Taoist temples in China today (Figure 6.4).

All such activities modify the microenvironment in significant ways: laid-out food and the burning of incense affect insects and small mammals. Pictures and statues accumulate dust, particularly in the interstices of Buddhist and Taoist shrines with their complicated carvings and accumulations of offerings and ritual objects.

Difference over birth control can have an environmental effect. Groups which abjure it such as the Amish and Orthodox Jews see their numbers rise although the effect on the latter is reduced by their ritual periods of abstinence and the limited time when sexual intercourse is permitted. The Shakers, by prohibiting sexual activity, dwindled in numbers almost to disappearance.

Differences in diet influence food production and waste. By the time children become adolescent, their diet pattern will follow that of their

Figure 6.3 Buddhist shrine for ancestors in Chiang-Mai, Thailand

parents for many years and will seldom be questioned. Some Hindus and Buddhists are vegetarian. Moslems and Jews avoid non-halal and non-kosher foods respectively, while Christians generally have fewer scruples. The Jewish use of alcohol in association with family meals leads to a different approach to drinking from that of Christians who themselves differ over whether or not to use alcoholic wine in Holy Communion.

Figure 6.4 Candles burning at Buddhist shrine, Beijing

Monasteries, convents, rectories, clergy houses and temples are home to religious personnel. Although these people lead dedicated lives and may be on call in much the same way as medical personnel, they have off-duty areas. Those who retire usually try to stay in the institutions where they have spent most of their lives. Even the cell of a Cistercian monk probably has some personalized characteristics which make it 'home' for him. Such homes, often adjacent to or part of churches, synagogues or temples, generate environmentally significant activities.

Popular shrines block traffic, as do 'saving the trees' movements under the impetus of particular Buddhist monks (Figure 6.5). Christian pilgrimages start and end at churches whose religious staffs have usually initiated the provision for the relief of stress and support for the poor and disadvantaged.

During the Black Death, the mortality rate of the parish clergy was substantially higher than that of monastic personnel who had no

Figure 6.5 Buddhist ordination of trees in Chiang-Mai, Thailand

parochial obligations or need for outside contacts. It is said that the two clergymen of different denominations in Eyam, Derbyshire, saved many lives in 1665. Both influenced their parishioners not to fly from the plague appearing in their village to avoid spreading the infection (Daniel 1966); 260 people out of 360 died but neighbouring areas were saved.

6.7 The use of foods

Food choices often derive directly or indirectly from religious guidance. Generally it is the better off who can afford to make significant variations in what they eat. Although not framed with any particular concern for the environment, these rulings affect agriculture, water use and waste disposal, whether they are part of a specific religious law or merely accepted by a particular group. There are also more complex correlations between regular diet and both morbidity and mortality which will affect human pressure on the environment.

Animals are widely used as food. Moslem and Jewish communities eat cattle and sheep but not pigs. Hindus eat the pig not beef. Moslem and Jewish communities have no pigs while in China they are especially numerous. In the highlands of Burma among Buddhists their abundance has had a significant influence on health (Tanner 1995, 85–94). On the other hand the same people tolerate a multiplication of stray dogs.

Breastfeeding by mothers is often a religious maxim or even an injunction. Its neglect significantly correlates with infant mortality, as does its duration with the delay in the resumption of menstruation and possible reimpregnation. Only the more economically developed societies have special foods for children. Elsewhere, the transition from breast milk to adult food is a particularly dangerous time.

Islam encourages the mother to breastfeed for two full years (Koran 2: 233) and ordinary Moslem women in India and Pakistan probably breastfeed their children for an average of 21 months (Potter et al. 1964). In polygynous cultures breastfeeding women do not have sexual relations with their husbands. With monogamy there is pressure for sexual relations to be resumed as soon as possible. The Roman Catholic Church supports this (Fromageau 1733) to discourage infidelity in the husband. In seventeenth-century France the need to abstain from sex encouraged the employment of wet-nurses which raised the birth rates of those women who did so (Lachiver and Dupaqier 1969, 1399). These alterations in birth rate influence the environment through their effects on agricultural demand and housing.

Many religions recommend fasting but to have a significant effect on the environment, it needs to be repetitive and widespread. Then it can reduce the need for agricultural production. Anorexia may have been a feature of early Christian saintliness. This is too individual to have any measurable effect on the environment since it would have been seen as too extreme for widespread emulation. Fasting as a penance seems to have been excessive when the confession of a single lustful thought brought six months on bread and water (McNeil and Gamer 1938), but confessions bringing such results were probably rare.

But religions do have required periods of controlled fasting. Medieval Christianity had long periods of obligatory and penitential fasting. Many Christians abstain from meat on Fridays and give up certain foods for Lent, although Vatican II has removed this as a canonical requirement. The Friday abstention, particularly in medieval times, may have been correlated with meat shortage and the difficulty of keeping livestock over the winter. A combination of these factors may have led to the creation of fish ponds in Britain to provide fresh food throughout the year and for special occasions. The contemporary Ethiopian Church has fasting up to midday and no animal protein except fish for 110–150 days each year (Knutsson and Selinus 1970, 956–9) and a total prohibition of camel meat (Simoons 1961, 108). This has bad consequences on young children and produces a clinically distinct form of gastritis from the eating of too much unleavened bread during fasts and overeating afterwards.

The outstanding example of fasting is the Moslem month of Ramadan. Since believers eat at night, the effect is to alter the daily times of food provision and removal. It is in some ways similar to the experiences of long-distance air passengers and crew, but probably less disturbing as it is associated with specially arranged nocturnal amusements, and there are permitted exceptions for the old, young and sick as well as pregnant and nursing women.

Vegetarianism is widespread in Buddhism, Hinduism and Jainism and in Hong Kong there are vegetarian halls attached to certain shrines (Topley and Hayes 1968, 135–48). Because these religions are widespread, the environmental effects of their beliefs are considerable through reducing animal production, and probably improving the health and longevity of devotees.

For both Hindus and Moslems these dietary religious restrictions have been of great historical significance. The belief that the East India Company's army was issuing new cartridges, greased with fat from pigs and cows, which had to be opened with the teeth, was one of the precipitating causes of the Indian Sepoy rebellion of 1857–58. This resulted in widespread destruction of property and many deaths. Today insulting behaviour using either of these meats has often been the cause of inter-religious rioting in India.

Some tribal religions have food taboos. Each clan will have a certain animal which members do not eat, which is therefore naturally protected. The overall effect is to keep wildlife at a level which that particular environment can sustain. In central Australia the major totemic sites of the Aranda tribe, which forbids the killing of the red kangaroo, coincide with the most favourable habitat of that species (Newsome 1980, 327). They believe that an individual's spirit children may reside in the bodies of animals. They forbid hunting and gathering at these totemic sites. But because these sites are also the most favourable habitat of the red kangaroo the embargo secures its survival. These Australian examples show how religion can generate environmentally useful acts in the absence of formal legislation and supervision (Bennett 1986, 130–7).

Some particular foods can have disastrous effects on a community and this reduces the effects on their surroundings. The people living in the Kuru valley (New Guinea) developed a lethal degenerative disease of the central nervous system in women which was initially assumed to be hereditary (Glasse 1967, 748–54). It was later found that they were eating the cooked brains of their dead relatives after making them into a pulp with their bare hands (Gajdusek 1973, 376–83). With the decline of cannibalism and the transfer of concern for the dead to memorial feasting with pork, the disease declined in frequency. Much the same lethal consequences came from the practice of drinking poison by Holiness Church members in the southern USA which they believed would not affect those free from sin (Schwarz 1960, 405–29).

Fasting affects different economic and cultural groups in different ways. Its results will be more marked in poor Pakistani and Saharan nomads than in the rich Arabs of Dubai and Bahrain, and more marked among the poor Christians of Ethiopia than in their prosperous co-religionists in Western countries, and probably among Orthodox Jews in Israel compared with their wealthy and more secular cousins in Western cities.

Diet affects health and agricultural demands. Seventh-day Adventists demonstrate some of the effects of a healthy vegetarian lifestyle. This cannot be attributed to diet alone although they do not smoke or drink alcohol and most also abstain from tea, coffee and colas. These characteristics alter their demand for agricultural produce and waste removal.

But as they do not live communally, there are few visible large-scale effects.

A comparison between Belgian Benedictine monks on a mixed diet and Dutch Trappist monks on a more frugal vegetarian diet found the latter living longer and with lower blood pressure (Groen et al. 1962, 456–70; Groen 1964, 310–13; Keys 1961, 100–5). Both had a lower incidence of heart disorders, asthma and bronchitis but higher rates of sinusitis, migraine, arthritis and rheumatism than the general population (Mackenbach et al. 1993, 569–73).

A vegan diet based on Zen Buddhism consists of cereals and the avoidance of many fluids. This can lead to severe nutritional disorders in children (Roberts et al. 1979, 296–8). Such consequences are augmented by the claims that such diets cure all diseases and ensure longevity (Ohsawa 1965).

Women in all societies generally keep the food rules and taboos more carefully than men. This is because they spend more time at home and have fewer meals prepared under different, and indeed secular, conditions. They are also less likely to suffer from pollution. On the other hand, Bengali women in Britain are reluctant to go outside for a variety of religious and secular reasons. This makes them likely to suffer from a lack of exposure to sunshine. The result is that Bengali men and women tend to receive pollution from the environment in different ways, men through vehicle exhausts and industrial wastes, women more through domestic smoke, ash and cleaning waste.

6.8 The environmental effects of communal religious buildings

Special buildings were a late development for almost all religions, but most are designed to last. Some, like the Buddhist pagodas in South-east Asia and the cathedrals of Europe, are subject to constant additions by new memorials to the dead and endowments for maintenance as well as for prayers for particular causes.

Even with the ability to build larger churches, religious groups may opt for smaller ones. The Cathedral of St George, the principal church in Addis Ababa, is no bigger than a large English parish church and relegates the 'less righteous' to the courtyard, denying even this to menstruating women whose devotion is limited to kissing the gateposts.

Many communal religious buildings are courtyard structures in which the control centre is indoors while most activities happen within the external enclosed area. The courtyards of the monastic church of Debra

Libanos overlooking the Nile gorge in Ethiopia, the major mosques in Delhi and the conglomeration of pagodas at the Shwedagon in Thailand and Myanmar are of this type (Figure 6.6). While the actual buildings and the courtyards may be swept rather than cleaned, the immediate environment outside the walls often has piles of rubbish and the body wastes of devotees for whom until recently there has been no provision. Catholic as well as Buddhist shrines can often be distinguished by the

Figure 6.6 Pagoda at Chiang-Mai, Thailand

piles of discarded packages of candles and incense sticks, rotting flowers and other offerings picked over by passing flies, dogs and birds.

Some communal religious buildings have added secular uses. Christian cathedrals have gift shops and all the main temples in Bangkok and even the mosques of Xian in northern China have a variety of shops selling religious artefacts, literature and general tourist paraphernalia. Buddhist and Taoist temples often have restaurants within their compounds in which lay visitors as well as the monks eat under conditions which may not be epidemiologically hygienic. Many Protestant cathedrals have restaurants attached to them and most parish churches and synagogues have annexes with cooking facilities. Some have crypts in which the poor can be housed nightly and fed.

While monastic buildings have always contained lavatories and wash places, no ancient churches or crypts had lavatories built into their designs. This caused pollution, especially to the water supply, around places of medieval pilgrimage such as the tombs of St Thomas à Becket at Canterbury and of St Cuthbert at Durham.

Burials have tended to be near religious buildings or sites. Church interiors mainly contained the bodies of those elite who could afford to pay for their interment near to religious relics or important symbols, or whose burial there was politically important. Much of the southern end of the Mount of Olives is covered by a large Jewish cemetery densely packed with tombs. There are notable exceptions. The late King Abdul Aziz of Saudi Arabia, a member of the austere Wahhabi sect, was buried in the desert in an ordinary cemetery indistinguishable from other graves with plain rocks at its head and foot. Many ordinary Christian dead have not been buried individually but in charnel pits or graves which were emptied and reused.

All religious buildings have a significant microenvironment sometimes containing bats, birds and a variety of insects. The wooden pews in the Roman Catholic cathedral in Merida, Mexico contained bedbugs in their cracks. Towers may provide protection for unusual animals such as the escaped eagle owl which lived for some time in the belfry of St Albans Cathedral in 1998.

Many places of religious pilgrimage have artistic and cultural attractions which attract secular tourists. This is especially noticeable in centres such as Assisi, Rome and Bangkok. Sometimes, old churches and temples lose their religious purpose but draw in large numbers of visitors because of the splendid architecture. The ruins of the ancient abbeys of Fountains, Rievaulx, Bury St Edmunds and Reading continue to attract tourists, perhaps sometimes as many as they had as centres of active Catholicism. The environmental protection of these places which was once ensured by their religious use is now continued more effectively by secular law. Thus many pre-Reformation sites which have lost their religious purpose have retained at least some part of their original environment.

6.9 Outdoor shrines and the environment

Lack of space in buildings sometimes compels religious activities to be held outdoors when season and weather are favourable. This includes processions, parades and rallies addressed by prominent speakers. The range of visual, as distinct from auditory, involvement was often large in meetings as diverse as Aztec pyramid ceremonies and John Wesley's open air meetings.

Many such gatherings centre on shrines. They are sometimes credited with helping people to cope not only with emergencies, such as droughts, floods and locusts, but also with the normal cycles of agriculture and animal husbandry (Binsbergen 1978, 48, 56). Outdoor shrines in large numbers are part of the scenery in many cultures in which they sometimes have great significance. Outstanding examples include the Ka'aba stone in Mecca, the grotto at Lourdes, and perhaps the Nazca lines in the Peruvian desert. Their environmental significance lies in the fact that they are out of doors and so can involve many more people on more occasions than could be contained in buildings.

Shrines may be natural or artificial. They can be classified into the following.

Natural shrines

1. Shrines created from or developed from natural features. These include mountain tops (Figure 6.7), springs, caves and groves. Their uniqueness and importance for devotees make them a focus of communal interest and concern. The sentiment holding together the marginalized Ik of northern Uganda appears to be the mountain behind their villages which is associated with the rain on their fields (Turnbull 1994, 169). In central Africa such outdoor shrines are occupied by spirits and have guardians who have traditionally controlled their surroundings. These guardians have advised when to burn grass and the order of the different operations in clearing the forest. They were therefore valuable in keeping the community in reasonable balance with its environment. Somewhat comparable

to such shrines is the annual decoration of wells and springs with complicated flower designs in Derbyshire villages which involves the whole community.

2. Shrines sculptured out of bare rock (Figure 6.8) or in caves which become the centre of devotional practices. Examples are the Coptic

Figure 6.7 Roman Catholic blessing of mountain peak

Figure 6.8 Myanmar. Buddhist pagoda on boulder

churches at Lalibela in Ethiopia, images of Buddha in Sri Lankha or the sites of papal masses as at Medjugorje.

3. Shrines around believed spiritual apparitions that are thought to confer healing, such as that of the Virgin Mary at the spring near Lourdes. This has transformed the locality with buildings (Figure 6.9) and enlarged the economic and social life of its community. The spring has no particular property which would explain the

Figure 6.9 Lourdes grotto with crutches of those miraculously cured

phenomenon or differentiate it from thousands of other springs along the flanks of the Pyrenees.

Artificial shrines

1. Shrines associated with politico-religious leadership. These are usually large buildings, arches, statues or tombs located in major cities. They include sites like Westminster Abbey, Les Invalides in Paris and the mausoleums of Mao Tse-tung in Beijing and Ho Chi Minh in Hanoi. The whole Forbidden City complex in Beijing was for the display and performance of rituals related to the environmental welfare of China as a whole. On a less grandiose level there have been many shrines where tribal chiefs through rituals have arranged or ensured the environmental welfare of their people. This group also includes village or neighbourhood shrines in which a smaller area than a chiefdom or state is connected to more specific spiritual forces. Examples of this are the spirit shrines outside Kachin villages in northern Myanmar.
2. Festoons of anonymous spirit invocations or thankofferings hung on trees or between posts, such as can be seen around the Potala Place in Lhasa and other monasteries in Tibet. More formal requests on wooden plaques are put on to prearranged racks near the Meiji and other temple shrines in Japan.
3. Shrines aimed to protect particular houses or landholdings. Notable examples are those guarding particular rice fields, the hex signs on the outsides of Amish houses, and the much-used spirit shines in houses and even on roofs and verandas of apartments in Thailand and Tokyo. Many Sukuma houses in north-western Tanzania have spirit shrines dedicated to particular ancestors which are erected in fulfilment of a vow but not maintained thereafter (Tanner 1969). Ancestor propitiation ceremonies take place at these shrines which also involves the dedication of animals to particular spirits.
4. Wayside shrines made by individuals to bless the traveller. There are many crucifixes and images of the Virgin Mary along roads in Catholic countries such as France and Italy. Of similar purpose are the small pagodas visible almost everywhere in the Irrawaddy valley of central Myanmar. These are erected as reminders of religious beliefs rather than places of regular worship.
5. Shrines commemorating the religious history of a region which are visited as a memorial act. Examples are the visits of Jews to the Wailing Wall in Jerusalem (Figure 6.10), places featuring the life of Christ visited by Christians, and the location of Mohammed's stoning of the devil, visited by pilgrims on the Hajj.

Figure 6.10 Wailing Wall, Jerusalem. Jews praying

6. Modern memorial shrines at places of personal bereavement as on the sites of road or rail accidents or graves. Mourners usually put piles of flowers and written messages at places associated with the disasters. A notable recent example was the vast number placed outside Kensington Palace and many other places in commemoration of Princess Diana.
7. Ambulant shrines which are taken out and paraded on special occasions (Figure 6.11). Christians in some areas follow large decorated floats on anniversaries. Hindus and Moslems stage competitive parades in many parts of India. In Manila hundreds of young men haul a heavy chariot carrying a statue of the Virgin Mary through the streets. Some people throw towels to the guards standing on the chariot to wipe on the face of the statue which are then thrown back to their owners with its blessing. These parades involve large numbers of people and huge expenditures of energy.

6.10 The environmental effects of religious cities

Many cities originated as religious foci. When these coincided with route centres and surpluses of marketable goods, they grew large. But many still retain their religious cores. In Europe they usually contain

Figure 6.11 Buddhist mobile shrines used in holy day parades in Sri Lanka

cathedrals with towers and spires. These enormous buildings in crowded medieval cities could rarely be seen as a whole and so the spires impelled passers-by to look upwards and reflect on spiritual things.

Some cities, such as Canterbury and St Albans, have retained an important religious function to the present day, despite the dominance of commercial activities. Others, such as Delhi and Mandalay, became political as much as religious centres, although are still known for important mosques and temples. Jerusalem was an ancient strategic centre in central Palestine. Since its capture by David in about 1000 BC it has had a continuing religious role, first for Judaism, then Christianity, then Islam. It has been the focus of many subgroups of these faiths, sometimes cooperating with each other but sometimes mutually suspicious and even hostile. All affect the environment through their buildings and such practices as baptisms, services and processions.

In pre-Islamic times Mecca was a centre where visitors, Arabs and non-Arab, were able to trade in safe detachment from the dangers and obligations of their own societies (Turner 1974, 231–70). Mohammed destroyed its pagan shrines and made it the centre of the new Islamic faith and of the Hajj pilgrimage. This pilgrimage confined to a single lunar month was manageable when the numbers were held down to some 40 000 by the sheer difficulty of getting there over desert tracks, especially when the lunar calendar brought Ramadan around to the hottest time of the year. Its subsequent central role in Islam has made it a privileged trading area free from the feuding which has afflicted the surrounding Bedu population. The invention of steamships facilitated Hajj travel but imposed no need to modify the Mecca environment. But once aircraft were used for the package movement of pilgrims almost on a door-to-door basis, the Saudi Arabian authorities had to cope with an annual flow of more than a million foreign and perhaps an equal number of Saudi pilgrims. This necessitated not so much changing as recreating the environment, although it remained largely unoccupied for 11 months in the year. It has recently been the subject of a series of research studies in part to assist the Saudi Arabian authorities in coping with the annual pilgrim flow into a city in a near-lifeless environment of dry volcanic rock, with limited access to water.

The pilgrimage rite for new arrivals begins with a special service in the Sacred Mosque on the seventh day of Dhu al-Hijjah, the last month of the Islamic year. Activities continue on the eighth and conclude on the twelfth or thirteenth. When about 10 km from Mecca the pilgrim enters a state of holiness or *ihram* and dons two white seamless sheets, cuts neither his nails nor his hair and abstains from sexual intercourse until the pilgrimage is over. He enters Mecca and walks seven times around the sacred Ka'aba stone in the Great Mosque and runs seven times between the minor prominences of Mount Safa and Mount Marwa. In the next stage he goes the few kilometres to the plain of Arafat for an afternoon and hears a sermon. He then returns via Muzdalifah, sacrifices an animal, and goes to Mina where he throws three stones at each of the three pillars there. The Hajj ends with a farewell circling of the Ka'aba before leaving the city.

In effect, modern Mecca is the desert equivalent of a greenfield site in which all planning has to be based on a maximum utilization of about 2 million people as against a regular use of perhaps less than a tenth of this number. It has involved building housing for pilgrims, and roads for vehicles and pedestrians between Mecca and the sacred sites. It has also required the provision of water supply, sewage disposal, electricity, hospitals and mortuaries. It is as good for Moslems to die at Mecca as for Hindus to die at Banaras. It has also required police and administrative staff, and slaughter houses and cold storage facilities for the enormous numbers of animals sacrificed annually. This quantity of meat cannot be eaten by pilgrims or even given away locally. It is necessary to organize its storage and ultimate disposal overseas. These facilities have been supplemented by a complete back-up of hotels, garages, souvenir shops and other businesses as well as the enlargement of the specifically religious buildings to cope with this flood of often emotionally excited pilgrims.

This particularized environment has in addition serious and unusual problems. The specialized medical needs of widely different foreigners has induced their home countries to provide their own medical teams to cope with the change from controllable epidemiological diseases to the present major causes of health problems, traffic injuries and heat exhaustion. The enormous tented camps require the disposal of waste from people perhaps not used to this type of life with its unfamiliar forms of social control. There must also be regular disposal of the great piles of stones that accumulate at the Mina pillars.

The whole organization of the pilgrimage to Mecca is an astonishing example of the quick creation of an exclusive and large religious environment which has no contemporary parallels. Even the religiously motivated concentrations of Mesoamerica, Egypt, South-East Asia and China do not parallel the overall complexity of its changes.

Banaras is thought to have been a religious centre since the sixth century BC (Eck 1981, 5). It has a role in Hinduism which reflects its location on the Ganges, a river which is seen as ritually sacred from source to sea. At Banaras there is a dedicated area surrounding the city which is occupied not only with many Hindu sects of varying esotericism in competition with each other but with a large Moslem population with its own obvious mosques. The numbers of pilgrims cannot be known as they come at all times of the year and there is no control point through which they must pass.

No vestige of planning has been or could now be applied except in peripheral matters such as attempting to reduce pollution in the Ganges, offering subsidized cremation, and putting flesh-eating turtles into the river to cope with bodies dumped into it. Not only is it a pilgrimage centre but, since it is a preferred place to die, its death rate is consistently higher than that of other cities in the region. It is also a common locality for suicides (Heber 1861, I, 162).

The city remains a religious focus within which there is a complex of integrated and self-sustaining religious groups. It would be almost impossible to achieve any overall agreement on environmental issues because each group is autonomous and actively defends rights over its territory. It is estimated that three-quarters of Banaras' existing population of more than half a million is dependent in some way on the pilgrimage trade (Saraswati 1975, 45).

The Vatican State, a small Roman Catholic enclave centring on St Peter's in the middle of the Italian capital, is the chief of the five basilicas in the city and is both a pilgrimage and administrative centre. Although it is their religious focus, Roman Catholics are dispersed over the whole city area and suburbs wherever religious houses, hostels, churches and hotels are located. Their religious impact is dispersed through the socio-economic and political life of a capital city, including its vast numbers of both secular tourists and religious pilgrims.

Finally, Salt Lake City, the present centre of the (Mormon) Church of the Latter-Day Saints, has a population of more than 200 000. It is the focus of a hinterland largely inhabited by people of this faith. Their impact on the environment is considerable because of the way the faith encourages church building, a healthy and abstemious lifestyle, practical education and high-quality farming. It is probable that its traditional polygyny has contributed to population growth.

6.11 The routeways used by religious practitioners

There is some difference between centripetality and centrifugality in the religious experience of pilgrimages. Centripetal pilgrimages concentrate in a single centre such as Rome, Jerusalem or Mecca. Movement to such centres fulfils the human urge to sense the wider fellowship of a great movement, to meet fellow pilgrims, and to assist in central religious ceremonies on an international scale. Participants fervently believe that the united prayer of such gatherings can influence divine action and that this can have a significant impact not only on themselves but also on the external community. Centrifugal pilgrimages, by contrast, seek remoteness and retreat from the world. Both kinds temporarily impoverish the pilgrims' home areas and enrich those visited. The environmental effects of the two types differ.

Centripetal pilgrimage will concentrate material changes into the major centres notably by demanding large religious spaces and buildings and by multiplying the requirement for food and housing. Most pilgrimages are of this type and concentrate environmental effects along routes and at specific destinations. They date from ancient time, as Semple wrote:

> Herodotus tells of 700,000 Egyptians flocking to Bubastis for the festival of Diana. . . . As early as 386 AD Christian pilgrims flocked to Jerusalem from Armenia, Persia, India, Ethiopia, and even from Gaul and Britain. Jerusalem gave rise to those armed pilgrimages the Crusades, with all their far reaching results. The pilgrimages to Rome, which in the Jubilee of 1300 brought 200,000 worshippers to the sacred city, did much to consolidate papal supremacy over Latin Christendom. As the roads to Rome took the pious wayfarers through

> Milan, Venice, Genoa, Florence, Bologna, and other great cities of Italy, they were so many channels for the distribution of Italian art and culture over the more untutored lands of western Europe.
> (1947, 99)

In the same category are the religious fairs and fiestas which occur frequently among all religious groups. They include the Hindu Kumbh Mela and the Moslem religious holidays, especially the 'Eid el-Kebir' or Great Festival in the last month of the Moslem year and the 'Eid es-Saghir', or Minor Festival, following the end of Ramadan. They also include the Anglican pilgrimage to Walsingham, and the Roman Catholic pilgrimages to Czestochowa and Santiago de Compostela and frequent saints' parades with decorative floats in southern Europe and Latin America.

The more important the religious destination, the larger will be the geographical area it serves, the more intense the use of routeways around it, and the more marked and the longer lasting the environmental effects. At one extreme in the British Isles is a medieval pilgrimage route near Carmarthen in Wales which has a cross cut into the rock wall where a granite block has been worn away by kneeling pilgrims. At the other extreme is Knock in Ireland (Coyne 1948) where an international airport has been created for pilgrim use.

Centrifugal pilgrimage, by contrast, disperses destinations to remote areas which usually require exceptional transport facilities. It follows an ancient eremitic tradition seen in early Christian monasticism as well as Oriental faiths. It reflects the desire of humans to 'recharge their spiritual batteries' and to seek religious purification by retreat to remote areas and sites. It is illustrated in literature by the quest for the Holy Grail by King Arthur's Knights of the Round Table, and in modern context by the retreat to ancient Christian sites such as Iona, Hindu pilgrimages to the source of the Ganges, and Buddhists to the temples of Angkor.

Before the days of aircraft, the great effort required to reach a destination discouraged many from trying. Some started but never arrived. Many of the earlier pilgrims from West Africa stayed on at Mecca and Cairo for years. Some had to earn enough money en route to enable them to go further and indeed to return home.

Some pilgrims make repeated attempts to walk to Mecca. In the past pilgrim ships used to dump pilgrims on the shores of Arabia and point them in the direction of Mecca, hundreds of kilometres away. The long walk would take weeks if not months and it seems likely that many have died crossing the desert. The junior author remembers camping near

Jebel Dhanna in the United Arab Emirates in 1965 when an Afghan pilgrim walked past on his way to Mecca, refusing even an offer of water. He told us it was his seventh attempt to get there after having been repeatedly turned back at frontiers. He had been dropped from a ship on the Emirate coast and still had over 1200 km of almost waterless desert to cross to reach Mecca. He had clearly devoted years to the attempt.

Although pilgrimages to major sites, however, often covered long distances, the majority have always been short and within national borders. Almost every country has a network of pilgrimage sites criss-crossing its surface such as India (Bhardwaj 1973) and Morocco (Eickelman 1976), as well as in Christian countries generally (Turner and Turner 1978). In Latin America there are more than 900 pilgrimage centres (Nolan 1991). There are also a growing number of non-Roman Catholic pilgrimage centres in Europe (Nolan and Nolan 1989).

In purely physical terms the passage of pilgrims alters the environment significantly. The passing of feet and hooves wears away the route and its immediate environment. Travelling pilgrims have often deforested and overgrazed the margins of popular routes. They have done this by gathering fuel for cooking and watch fires and fodder for the animals to support themselves and carry the well-to-do, the elderly and the sick.

Damascus to Mecca took an average of 40 days with no halts of more than an hour, then three weeks in Mecca to recover and for trading and then return. The French observer Volney who witnessed the 1787 convoy estimated the usual number of pilgrims at between 30 000 and 50 000 (Rafeq 1966, 61). This imposed a considerable strain on the environment along the route and at Mecca to where all supplies had to be brought in by camel.

Some of the inhabitants of the desert saw these Meccan caravans as an annual source of income through robbery and extortion. Handbooks on the Hajj written for use at the end of the nineteenth century warn against dismounting before the caravan is secure, urge the need to sleep in shifts in order to protect the camels and personal property, and never to leave a sleeping place in order to relieve oneself (Kazemzadeh 1912).

Pilgrimage routes have always been epidemiological channels along which diseases pass into new areas and new infections are brought back. Contagious diseases result from close proximity and such means as carrying of germs in samples of special waters and the communal kissing of sacred objects, which can give mouth-borne diseases. Both may carry infections to distant families. Pilgrimage-related epidemics have reportedly often claimed well over 100 000 lives. A Madras cholera epidemic started by pilgrims in 1875–77 killed no less than 350 000 Hindus (Bhardwaj 1973), and in 1974 300 Nigerian Hajj pilgrims died from cholera. In 1831 cholera is estimated to have killed half the pilgrims and resident population of Mecca (Pollitzer and Swaroop 1959). Deaths from cholera related to the Kumbh and Ardha Kumbh religious occasions in central India peaked at 178 000 in 1894 but still killed 52 000 in 1948 (Banerjea, 1951, 17–46). The Kumbh Mela in 1989 is estimated to have been attended by 10 million pilgrims. Their passage along the railway routes to their homes could be traced by the outbreaks of disease. The Sudan was consistently infected by epidemics of relapsing fever and smallpox through the passage of West African pilgrims en route to Mecca (El Tayeb 1976).

A Sunday visit to the shrine of Our Lady of Guadalupe near Mexico City illustrates the health problems involved. Some people move on their knees from the gates to the church on occasion leaving blood on the gravel. The crowds in the church are dense especially when they place piles of flowers along the altar rail and take communion from the hands of the priests. The high temperature as well as an absence of lavatories add to the health risk.

Polluted water is a health hazard especially in large pilgrimage centres such as Lourdes, Banaras and Mecca. At Lourdes the sick are bathed in water which is not changed after each sick and possibly incontinent pilgrim has been lowered into the baths (Marnham 1980). Most pilgrim areas at peak periods have serious difficulties over fouling from human effluent. Even where public toilets and washrooms are available not all pilgrims can use them.

But religious routeways also transmit ideas. The Crusades illustrate this in the diffusion of Western Catholicism in the Holy Land. The impact was increased by the work of the religious orders: the Knights of St John of Jerusalem (Hospitallers), who were founded in the eleventh century to care for pilgrims, and the Knights Templars, founded in the twelfth century, to protect them. Both acquired castles and maintained permanent garrisons in Palestine at the time of the Christian Crusader kingdoms. The Hospitallers maintained castles first in Rhodes and then in Malta, and the Templars had centres throughout Europe in the Middle Ages.

It has been suggested that the cultural identity of India is due mainly to the grand pilgrimage recorded in the *Mahabharata*, a popular Hindu epic, which covered the whole subcontinent (Eck 1981, 336). In ancient times pilgrims made these long journeys often enough for them to be recorded in legend. They would have visited Kashmir, Banaras, Mount Abu and Kanya in the extreme south of India, involving them in about 2000 miles (3200 km) of travel. The average distance travelled by pilgrims going to Badrinath, the source of the Ganges was 490 miles (780 km) and to Hardwar and Kangra over 170 miles (270 km) each way (Bhardwaj 1973, 118).

Islamic Morocco is also criss-crossed by well-used routeways to the tombs of popular saints such as that of Moulay Idris. Pilgrims circumambulate the shrine as other Moslems do around the Ka'aba in Mecca on the Hajj (Crapanzano 1980,16–17). They consider the dead saints to be still in residence, bathe in waters sacred to them, and remove handfuls of earth from the sanctuary dedicated to them.

Pilgrims create service industries which manufacture, distribute and market religious goods. The Taiwanese pilgrims within the Formosan circuit of sites fill their bus with secular and religious purchases and even complex pamphlets on Taoism which they do not have the necessary calligraphic knowledge to read (Sangren 1987, 193). Practically, the sites become religious parks in the same way that many cities have industrial parks in their suburbs. Such modern religious foci as Fatima, Lourdes and Medjugorje are at the centre of extensive developments marked out by new buildings, roads and car parks.

6.12 The natural environment and sacred spaces

A sacred space is not normally extraordinary in its characteristics. It is made so by religious choice. The bo-tree at Bodha-gaya is no different from any other specimen of *Ficus religiosus*. Mecca was an undistinguished trading centre and a neutral area free from tribal fighting at the time of the Prophet Mohammed's inspiration. This indeed may be one of the reasons that it developed into the annual Islamic pilgrimage rather than Jerusalem.

Sacred spaces have come into being in three types of localities: outstanding natural features, commemorative sites and utopias. Some outstanding natural features have attracted human imagination and evoked a religious response. This especially applies to vertical symbolism with heaven above and hell below. Thus the esteemed features are often mountains. There is a tendency to see the spiritual world as above and the material world below, with the mountains connecting the two.

> People in widely different parts of the world regard the mountain as the place where earth and sky meet. It was the central point, the

> world-axis, the place impregnated with sacred power where the human spirit could pass from one cosmic level to another.
>
> (Tuan 1974, 70)

Thus it is understandable that the elevation, remoteness and beauty of some mountains have imbued them with sanctity. By contrast, however, some cultures have conceived their ancestors as emerging from the earth below.

In Ireland the summit of Croagh Patrick has been the focus of devotion for many centuries with mass being celebrated on the summit; 40 000 Roman Catholics performed the annual pilgrimage there in 1972 (Turner and Turner 1978, 208). In Sri Lanka the prominent mountain Adam's Peak is the centre of pilgrimages by all four of the island's religions. This involves several hours' climb to the footprint at the summit which was made by the Buddha, St Thomas, Shiva or Adam according to Buddhists, Christians, Hindus and Moslems respectively.

The Masai of Tanzania call the volcano Ol Donyio Lengai the 'Mountain of God'. Both Moslems and Jews ascribe a special sanctity to the Temple Mount in Jerusalem. Mount Olympus in Greece and the four mountains in China are the object of pilgrimages. T'ai Shan, the chief of the latter, was a divine creation where the Emperor Wu (140–87 BC) sacrificed to heaven and earth (Chavannes 1910). Such areas are not always the most distinguished geographically. Mount Everest is not an object of particular veneration although it houses a spirit according to Buddhists and is considered sacred enough to have taboos on sexual behaviour there. Mount Kailas and Mount Abu are much smaller Indian mountains but are sacred to Buddhists, Hindus and the latter particularly to Jains.

Societies which have no convenient mountain near by but possess the necessary resources erect facsimiles to connect with the sacred. In Mesopotamia the belief that a mountain united heaven and earth, led them to build ziggurats at Babylon, Ur, Erech and Asshur. These were visible from afar and represented cosmic mountains (Tuan 1974, 70).

Hindus have the mythical Mount Meru standing at the centre of the world. The Navajo, Tewa and Zuni American Indians have their territories bounded by mountains which are more mythological than geographical features. The Ik of northern Uganda, who are recorded as having virtually no religion or social structure, seem to derive their spiritual consolation from staring at the mountain dominating their uncertain environment (Turnbull 1978, 197).

Circumambulations are sometimes a form of religious observance. Nepalese and Tibetan Buddhists have used their mountains and other sacred topographic features for specific rituals. They walk around them not simply as a physical exercise for spiritual well-being, but as a productive activity that creates specific amounts of virtue, furthering the individual's progress towards Liberation. Should the pilgrim be carried by another human or animal, then this increased virtue is shared in equal proportions with those who carry them.

The easiest of such circumambulations are around living lamas, images, volumes of the scriptures, offerings and monasteries, while those around mountains and lakes are the most strenuous. The circuit around the Koko Nor lake in Tibet is 269 km. The circuit around Amdo, another Tibetan sacred mountain, takes seven days and devotees can increase their merit by doing the circuit by a series of caterpillar-like prostrations along the track (Ekvall 1964b, 226–50).

The second type of sacred or semi-sacred localities is shrines of national or regional remembrance such as war memorials and cemeteries. Europe has many square kilometres of war cemeteries. Americans maintain Arlington National Cemetery and have preserved the submerged hulk of USS *Arizona* sunk by the Japanese in their attack on Pearl Harbor as a visitors' centre for tourists, veterans, survivors and patriots (Chidester and Lidenthal 1995, 4). The eternal flame over the tombs of the unknown soldiers in different countries who died in the First World War, the memorialization of concentration camps, and the Holocaust memorial in Israel, are also of this type.

The third group includes sites which enjoy a special divine interest or presence. In Israel this concept attaches to the territory of Palestine as a whole. In Greece, the Church ascribes a special significance to Mount Athos, whence it excludes all women. The Latter-Day Saints see the area around Salt Lake City as the promised land. Their earlier attempts to create a Utopia at Carvan led to their violent expulsion and westward migration. Jim Jones of the People's Temple attempted to create a spiritual Utopia in Jonestown which led ultimately to the mass suicide of all 900 inhabitants. A corollary is that the exclusion of outsiders from such locations can provoke their hostility. This is seen, for instance, in the resentment of Outcastes in being barred from Hindu temples and in the historic conflicts between Christians, Jews and Moslems over access to the holy places in Jerusalem.

What principles underlie the existence of these sacred spaces (Lane 1988, 11)? Do worshippers choose particular ones for their devotions because of people or events attached to it, or do their religious beliefs predetermine such attraction? This raises the whole question of the conditionality of historical events. Did Mecca choose the Prophet or the Prophet choose Mecca for relaying God's messages? There is an element of truth in both. The Prophet chose Mecca but this was because providence had determined that he live there and that his wife's business was located there when he received his inspirations. Likewise was the Buddha's Enlightenment under the bo-tree at Bodha-gaya due to the appeal of the environment or does his inspiration give the setting its significance? Both the person's thoughts and the surrounding circumstances are needed, but the combination could be viewed as itself a result of a deeper providence.

Some sacred places have no clear initiating human factor. The piles of stones at mountain passes in the Andes and Himalayas or near promontories on Lake Victoria in Tanzania, must have been started by someone but did not give rise to a cult and the originator is unknown. Much the same would apply to trees festooned with petitionary messages or strips of cloth surrounding the Meiji shrine in Tokyo, around the Potala Palace in Lhasa, Tibet, and beside roads in Cyprus and Sicily. An oak tree near Silbury Hill and the West Kennet Long Barrow burial chamber in England was recently festooned with ritual offerings.

The permanent residents around these centres often do not themselves find them spiritually helpful. They may prefer to visit others far from home. The people who live in Lourdes probably go to other sacred areas and may not be generous to the pilgrims. Some may exploit them in the same spirit that has prompted some Arabian Bedouin to rob the annual Hajj caravans.

The religious experiences which come from arriving in the sacred space are not related to the glories of nature or cultural aesthetics. Indeed outsiders are as likely to be put off by the created ambience surrounding these established sites, especially the rows of stalls selling religious trinkets. Whether at Lourdes or the Buddhist pilgrimage centre at Katagarama in Sri Lanka, there is little visible at the site to inspire the pilgrim.

Nevertheless, the vast majority of pilgrims come away from the sacred space religiously invigorated. This is the draw that brings millions to these places. It would be virtually impossible to find Moslems who have performed their once in a lifetime pilgrimage to Mecca, who do not feel that their visit has enhanced their private and public status as Moslems. The reason must be sought in the mind and heart of individuals rather than in the objective value of the places themselves.

7
Socio-economic Activities and the Environment

7.1 The impact of changing socio-economic conditions on religion

Religions react to changing socio-economic circumstances, political domination, and above all, to new ideas. These traditionally arrived from missionaries, traders and returning pilgrims but increasingly from books and most recently from the media. The resulting changes can be drastic and rapid as we have seen in the origins of Christianity and Islam.

The absence of written formularies can perhaps make non-literate religions more adaptable to societal changes. Their practice derives from traditions passed down by word of mouth. Their messages inherited from the past, even if manifold, can be couched in such general terms that they are capable of being applied to any number of situations. This applies particularly to Hinduism, Taoism and Confucianism. These qualities help give them a flexibility which enables them to keep pace with surrounding social and environmental changes.

Religions which have written scriptures must continually relate their dated statements to material cultures which did not exist when they were written. Vast societal change as well as widespread literacy, television and computers have altered the communication of religious ideas and multiplied their effects on the environment. Easy travel and powerful modern weapons have enabled the religious zealot to be far more environmentally damaging than he could have been in past centuries.

Judaism and Christianity are exceptional in the inspired selectivity of the details recorded about the lives of the prophets and of Christ himself. They have moulded the behaviour of Jewish and Christian societies ever since. Government, jurisprudence, finance and social mores all reflect them. The Koran and the Hadith are the bases of Islamic law and actions. The lives and teachings of religious leaders such as George Fox, Brigham Young, Mary Baker Eddy and Ellen G. White have strongly influenced the day-to-day behaviour of their followers.

Faiths vary in their ability to meet these socio-economic changes. The Roman Catholic Church, for instance, can adapt its teachings to meet changing needs through its teaching arm, the magisterium. The Pope, by canonizing certain individuals, makes it likely that their graves or places where they had significant religious experiences will receive the attention of pilgrims. Changing the rules on fasting on Fridays and rulings on artificial birth control have also had environmental effects. Islam faces a particular challenge in relating to non-desert environments in high latitudes and where business depends on monetary interest. Some faiths may have so exacting an original message that they cannot easily change without losing their identity. The Shaker faith is of this type. Its rules, framed in the early nineteenth century, separating the sexes from virtually any contact, make adaptation impossible.

7.2 Religion and specific occupations

Religious belief influences most secular occupations in some way, and through this, the environment. Many occupations have patron saints or connections with particular temples or holy places and ascribe success to their help or failure to their absence or opposition. Either result can have measurable effects on the environment. Many Ghanaian farmers had earth shrines to ensure good health to look after the crops at critical periods and to bring good harvests. These have lost support and given way to shrines to spirits which cater for a more mobile population and wider geographical spread than the village areas of most people (Werbner 1989, 225–6).

At one time England had some 40 000 monasteries, guilds and charitable institutions with such spiritual patrons. St Stephen was thought to look after stonemasons, St Joseph carpenters, St Anastasia weavers and St Andrew fishermen. St Giles was the patron of the sick, which accounts for the location of churches dedicated to him away from city centres.

Today, Bolivian tin miners use the Christian calendar as a framework within which they have to make sacrificial offerings to Supay, an older deity than any Christian saint. At the entrance to the Itos mine there is a niche for the statue of the protecting saint for whom masses are said. The distinction between Christian theology and the older paradigm of earned protection from some other worldly power was made clear by the sermon of the Roman Catholic priest at one such celebration when he said that the statue was not placed there to bring the miners good luck. Nevertheless the miners continued to make offerings to the saint to ensure their protection from accidents (Nash 1975, 140–1). Even in their village rituals Andean villagers express their overriding concerns for productivity and fecundity in terms of their pre-Christian gods (Isbell 1978, 138).

Occupations that involve danger often tend to have enhanced religious awareness. Iron smelting in small traditional hand-pumped forges and the throwing and firing of pots by hand are liable to accidents and lead their practitioner to seek other-worldly support. Few hunters search the forests for food without seeking the guidance and support of spiritual forces as exemplified by the Gindo of southern Tanzania in their search for wild honey (Cross-Upcott 1956, 81–109).

Religion often imposes constraints on the choice of employment. Some Hindu castes are consigned by birth to specific work. Most Moslem women are largely confined to their homes. Jews and Seventh-day Adventists avoid Sabbath work and non-kosher diets. Jehovah's Witnesses until recently would not accept employment involving or receiving inoculations. Christian Scientists abjure conventional medicine. Many Christians, especially Quakers, Plymouth Brethren, and Seventh-day Adventists, would try to avoid military service. Groups such as the Amish farm in ways that do not use petrol-driven vehicles. All these constraints have environmental consequences. Sabbath and Sunday observance, in particular, reduces the workaday exploitation and pollution of nature.

While the British regarded the Hindu caste system society prior to Indian independence as a rigid framework, it now appears that this structure was less so than they thought (Bayly 1999, 25). Hindu society functioned mainly in terms of innumerable largely occupational sub-castes whose members' social and religious positions depended on their degree of observance of local standards of relative religious purity and the local economic, social and political status of the sub-caste (Srinivas 1987, 82). There was no all-Hindu structure which could allow the member of any caste the same religious and social privileges everywhere.

The emergence and role of allegedly nationwide caste associations, paralleling European guilds, are not as important as had once been thought. Caste groups therefore may have been functionally limited to a particular village or area, such as the management of the cremation ghats at Banaras. Older housing areas often contain single sub-caste residential streets, many of which were also occupational areas for trades such as jewellers, or hamlets and slum enclaves for the Untouchables.

Earlier studies of the power of castes concentrated on agrarian, criminal and landless castes, but significant recent evidence has come from the Chettiar banking sub-caste of southern India (Rudner 1994, 84–7). This sub-caste had important environmental effects. At one time there were 1500 Chettiar firms financing the rice industry in Burma so that it expanded from controlling over 3 million acres (1.2 million ha) under rice in 1882 to 33 million acres (13.2 million ha) in 1932 (Furnivall 1956, 56–7). During the Depression in the 1930s, the Chettiars foreclosed on their clients and ended up owning 30 per cent of all rice-producing land in southern Burma with themselves as absentee landlords. This accounted for the build-up of hostility to Indians in general in pre-war Burma. When the Japanese invaded in 1942, they fled to India and thousands died initially from Burmese brutality but then from starvation and disease in the mountains along the border.

One Untouchable sub-caste has the big task of handling and disposing of thousands of Hindu corpses in Banaras. This work is so profitable that they defend their ancient rights to this unpleasant work against other sub-castes who attempt to encroach on their monopoly (Parry 1994, 57–69). With a daily throughput of at least 100 corpses, these men have a major environmental impact. Even with the presence of a subsidized crematorium, many bodies are only half-cremated because of a shortage of wood. These are added to the unburnt bodies already immersed in the Ganges. The recent decline in the estimated number of these unburnt bodies may only reflect a decline in infant mortality correlating with the increase in hospital confinements.

Christian monasteries are a notable example of religious impact on agriculture. Earlier foundations followed St Benedict in aiming to be self-sufficient in food and clothing (Coulton 1925) and to manage the earth so that it retained its productivity (Dubos 1974, 122–30). They attracted quite large numbers of monks as well as lay brothers who, possibly because of the harshness of secular life, worked on the monastic farms. The extent of monastic holdings shows their importance in the medieval European environment. By the fourteenth century the western Christian Church through its religious orders owned 20 per cent of English and 25 per cent of Castilian soil. It was because of this increasing monopolization of land that political opposition to ecclesiastical bodies began to emerge (Boissonnade 1927, 244).

The Cistercians carried St Benedict's call for self-sufficiency further and became large landlords. In AD 1216 their great abbey in Silesia owned 8000 ha there, 32 000 in Hufen and 48 000 in Netze. As their holdings increased, monks and priests became less and less involved with religious activities and acted more and more as controllers and managers.

Since the monasteries were often encouraged to open up wild areas, rises in the price of wool as opposed to cereals led them to keep sheep. But there were never enough men to farm the enormous acreages which they came to own. As living conditions improved the numbers of religious declined and the abbeys were left with enormous landholdings which the remaining monks were not trained to manage. This caused severe problems when wages and rents rose and prices declined (Postan 1966, 59). The high prices of wool made it more profitable to turn three-quarters of the abbey's tenant farmers off their holdings and to let the land go for grazing sheep (Coulton 1925, 225). They became an important part of the wool trade on which medieval England largely depended. In Yorkshire alone the remains of the five great abbeys of Fountains, Kirkstall, Rievaulx, Jervaulx and Byland show the extent of this trade before the Dissolution. They also sometimes failed to control their numerous free-ranging pigs. In AD 1250 the Abbey at Loos was legally restricted to 120 (Rosny 1837, 36). It is hardly surprising that there was little popular opposition to the abolition of the monasteries in England.

Landholdings shrank as monastic landlords were no longer able to farm all the land they owned. The Benedictine Abbey at St Germain des Prés owned at least 32 748 ha at the beginning of the ninth century which had shrunk to 2434 by AD 1384 (Ganshof and Verhulst 1966, 303–32). Much the same decrease in available labour faced the Jesuits in Zimbabwe who had been given 100 000 acres by Cecil Rhodes when it originally came under British suzerainty.

There were other reasons why monastic holdings decreased. Numerous donors in the Germanic kingdoms withheld their land during their lifetimes, thus securing provision for their old age and sickness as well as supporting those who went on pilgrimages. This system performed the functions which are now carried out by insurance companies and banks (Dopsch 1966, 200–1). There is scarcely a trace of serious manual labour by Christian monks after AD 1300 with the partial exception of the few Carthusians, and even they seldom worked with their hands outside their own cells and gardens (Coulton 1927). Surviving monastic account books show that they did no domestic work, and that they were often fewer than their servants. Romsey Abbey about AD 1270 had 57 servants and Bury Abbey in AD 1280 with 50 monks had 111 servants. So the monks had their main environmental influence through directing their employees and tenants (Dugdale-Caley 1817–30, 2, 549; 5, 365).

A somewhat analogous pattern of expansion and collapse occurred with the American Shaker sect in the mid nineteenth century. This avowedly Christian farming movement reached its peak in the decade before the Civil War with 6000 members recruited mainly from the poor and those in debt. They worked their farms themselves with considerable success and invested their surplus capital in farmland. This repaid them well while they had the labour to work it. By 1875 they owned 80 000 acres. Even when they could only cultivate a part themselves, they still bought land which they rented out or managed with hired labour, but this violated their basic principles of separation from the secular world (Andrews 1963). Just as the monasteries failed to recruit sufficient lay brothers to work their fields, Shaker membership declined as outside opportunities for personal advancement increased and land became available in the western states. Both monasteries and Shakers, therefore, became victims of their own success.

7.3 States and societies with one dominating religion

A religiously united community is in a better position to handle environmental problems and to control exploitation than a divided one. But much depends on its size. Small non-literate tribes often have a religion with practices which ensure a balance between the community and its environment. For instance, the Zuni society of hilltop villages and limited water has a religion concentrated on villagers' concern for rain (Bellah 1966, 231).

They also face the need to make doctrinal changes even if they do not change their faith. The Sukuma of Tanzania have a largely intact non-literate traditional religious system and a low-level involvement with Christianity. The traditional faith mainly involves communication with and propitiating ancestral spirits. But independence has brought economic, social and political changes. They have adapted to this by reducing the influence of ancestors and increasing the significance of free-ranging non-specific spirits, witchcraft and concern for the psychic malevolence of their neighbours.

As tribal societies extend into new areas they meet new challenges. The traditional religion of the Chagga tribe of Mount Kilimanjaro has been almost entirely replaced by literate Christianity, although this has not prevented a continuation of traditional religious practices. The tribe's conversion has contributed to a rise in population through improved health and education, causing it to expand beyond a limited area of temperate rainforest and to colonize further and further down the mountain into the drier plains below.

Where a dominant religion restricts minority faiths it reduces any environmental impact they could have. This occurs in Saudi Arabia where only Islam is allowed, but also in countries such as Pakistan, India and Jordan where Christianity is allowed to provide social and educational services but not actively to seek conversions.

Islam, because of its identification of political with religious power, can be psychologically and physically suffocating to minority faiths. Some nation states, such as the Taliban-controlled areas of Afghanistan, enforce a range of Islamic practices by political action. This has put power over environmental matters into the hands of a socio-theological elite with a perhaps over-intellectualized view of what can be achieved. It has had a significant indirect effect on the environment through making it difficult for women to get an education or work and thus excluding them from a public role in repairing that country's civil war disrupted economy.

It is practically impossible for any centrally organized religious system, administered by a religious police, such as in both Saudi Arabia and Iran, to ensure conformity or fulfil all their peoples' religious and psycho-social objectives. Subsystems will have developed in private in the same way that religion survived in Communist Russia especially within denominations that did not require specially trained and ordained ministers. Likewise in China the Falun Gong was able to organize itself through the computer Internet.

Excessive theocratic control can lead to opposition and even rebellion. Iran has an overtly theocratic governing elite whose retention of clerical and religious property rights and privileges has become a popular theme of discontent. Inequalities of wealth also involving the Church were one of the principal causes of the English Peasant Revolt of AD 1381 which succeeded in controlling London for a time, and in the popular support for Henry VIII's dissolution of the monasteries in the sixteenth century.

Attempts to impose the Islamic law against the charging of interest have been largely ineffective because they have seldom been enforced with great vigour and have always left ethical loopholes such as the repayment of loans with additional gifts. But there has probably been some environmental impact resulting from the discouragement of lending.

Human factors also modify the effects of interest embargoes in monolithic Moslem societies. Sir William Willcocks (1935, 273–5) recounts some experiences of the Egyptian Agricultural Bank set up to provide capital for farmers (*fellahin*). The loan recipients often spent the money not on agriculture but on acquiring second and younger wives. To permit this their older wives exacted the return of money from their original dowries. When the farmer subsequently tired of the younger wife and sent her away in order to return to his original companion of many years, she also exacted the return of her dowry on the same grounds. The result of this process was to transfer the assets of the Bank into the pockets of the *fellahin* women, who then made money by relending it, not to their husbands, but to other men in their village.

Other societies with official religions have been less monolithic than Islam. China in the past had and Thailand today has theological systems in Taoism and Buddhism respectively which eschew both centralization and religious conformity. The resulting lack of religious homogeneity has made it difficult to isolate specific environmental effects of religious practice in these countries.

Some consider that the unsympathetic introduction of Christianity can be unnecessarily socially disruptive. It has been argued that this applied to the situation which faced Australian Aborigines in their first meetings with white missionaries:

> Aboriginals called together to receive gifts including axes at a mission Christmas party found themselves facing one or two whites who sought to control their behaviour for the occasion, who disregarded the age, sex and kinship variables of which the Aboriginals were so conscious, and who considered them all at one subordinate level. The role of traditional ideas, sentiments and values were undermined at a rapidly mounting rate with no new conceptions being defined to replace them. The result was the erection of a moral void which foreshadowed the collapse and destruction of all Yir Yoront culture, if not the extinction of the biological group itself.
>
> (Sharp 1952, 457–64)

Thus, the introduction of a new faith, if it is accompanied by the technological and political superiority of its propagators, can damage or even destroy the environmental understandings of the people who accept it. These understandings may be integral to the former non-literate faith, giving guidance on how to meet the believers' material needs and ensure their survival. The new faith, in removing the 'spirits' related to particular geographical features and totems related to particular animals, runs the risk of dismissing the environmental understandings which are linked to them. The change may actually allow harm to the environment by leaving it open to exploitation. Belief in such spirits may, for instance, have protected water sources by preventing the cutting of trees around them. Replacing belief in these 'spirits' must be accompanied by an understanding of the conservation principles they may have served.

Most villages in the world tend to have agricultural lands around a nucleated centre. The original reasons were for security and to facilitate communal activities including religion. New England townships radiate from their meeting houses with the boundaries of grace at the edge of the surrounding forest (Lane 1988, 110). The straight lines in Shaker villages and even their furniture were shaped to a heavenly ideal (Sprigg 1975, 95). This geographical arrangement helped the church to play an active part in the agricultural year, following to some extent the Jewish festivals, with Easter equivalent to the Passover at the time of the barley harvest, Whitsun equivalent to the Feast of Weeks at the time of the wheat harvest, and the Harvest Festival and American Thanksgiving equivalent to the Feast of Tabernacles at the close of the olive and fruit harvest. Fallow years were also an echo of the Jewish Sabbatical years.

Some religions may be in a minority at national scale but dominate a region or a stratum of society. The Mormons of Utah are an example of the first, Hindu castes of the second. The Mormons exemplify modern land development under religious inspiration. They fled from the northeastern states to the West to avoid persecution and seek independence for their beliefs, which notoriously included a tolerance of polygyny. Under the leadership of Brigham Young they took over a barren and unoccupied area in Utah and turned it into a prosperous agricultural community by hard work and ingenuity. Brigham Young also instituted many programmes to achieve unity and facilitate migration and colonization. The object was to create the City of Zion on earth.

This project involved extraordinary tenacity and thoughtful management without outside help. It appeared particularly in the public ownership and administration of the water running off the mountains. Each man gave his labour to build the dams and ditches in proportion to the amount of water he needed for his farm. This was so successful that by the end of the nineteenth century the Mormons had provided a livelihood in this inhospitable region for half a million people, an extraordinary achievement which provided the foundation for a distinctive way of life (Arrington 1958, 52–62, 88–137, 411). Furthermore, the Mormon Church always sought to create a system which was visually pleasing and accorded with the biblical principles given in Numbers

35: 1–5 and Leviticus 25. These include injunctions to provide for the support of religious leaders, to fallow the land every seventh year, to enjoy occasional jubilees, to remit slaves and protect the poor.

Farmers redirect water from dependable distant sources on to their farms in strict rotation, both day and night, so that irrigation ditches are characteristic of almost all Mormon villages. They use the roadsides for grazing and hay. They have widely spaced brick or stone houses on large blocks of land with clusters of four at street corners rather than in the middle of their plots to encourage social contacts and mutual help (Francaviglia 1978). Even in areas in which the Mormons are only a small part of a community which contains people from four other cultures and religions such as Rimrock, New Mexico (Vogt and Albert 1966), there is a distinctive pattern on their holdings which sets them apart.

The Mormon system of working the land as a religious undertaking was successful even where individual farming enterprises would not have succeeded. The strategy was to populate vast areas of the American West with self-sufficient farming communities, socially and economically separate from American society. By the 1880s there were 20 Mormon farming settlements in the Little Colorado river basin but 20 years of poor harvests or complete crop failures meant that they only survived by subsidies. The church headquarters provided these under a complex system of resource redistribution based on the paying of tithes in kind from every Mormon settlement.

The Mormon leaders saw their new environment as providing a unique opportunity for circumventing local environmental limitations by adopting a multi-habitat exploitative strategy. Individual settlements could utilize the region's spatial diversity to counteract temporal variability. The hierarchical organization of the Church facilitated the concentration of surpluses for distribution to meet local needs through their own system of storage, transport and credit. Redistribution of church tithes effectively integrated the population and productivity of every occupied habitat. This worked much better than joint enterprises linking only a limited population within a few closely related habitats. Contrary to what agricultural economists might well have estimated, the Little Colorado Mormon settlements survived (Abruzzi 1987, 317–38). In this way the Mormons colonized a semi-desert through centralized planning based on a religious idealism and beliefs which many outsiders considered to be less than rational.

The Hindu religious system affects agriculture in very different ways from the Mormon. Its impact is more through societal strata than through regional dominance. Both traditional and current thinking in

India sees human relationships within an ideological framework that accepts caste. A person is born into a caste which is part of a hierarchy with the Brahmans at the top and the Outcastes excluded at the bottom. These are not usually separated geographically although in Kerala social distance between castes was traditionally expressed in spatial terms (Srinivas 1987, 47–92).

In practice the caste system is much more complex. There are not just four castes to consider but several thousand sub-castes. Each has complex local permutations as to who can eat with whom or take water from whom. Yet they have to live communally. There is thus an ambiguity in all social relations. Complex sub-caste loyalties surface in particular situations. This is full of potential for role conflicts but in practice is avoided by the accommodations of village life based on the ownership of land.

If a village should contain sufficient Brahmans to dominate its life, then there would be an overspill from this accepted superior religious purity into all other areas of their social and economic lives. If a lower caste dominates and has social and economic power, its members could invade the temple area controlled by the Brahmans. The latter would not be in any position to complain as their living depends on the contributions of others. A villager would not, however, see it as sensible to disturb the social balance for an ideological principle supported by the Indian Constitution.

There are religious factors which insulate sub-castes from each other. In Mysore, for instance, these include endogamy, occupational specializations, the distinctive traditions of each sub-caste and the ban on commensality, although this last is compensated for by donations of uncooked food when there are village celebrations. All of these practices are within the ill-defined concepts of pollution, karma and dharma.

But this is a village community whose specializations require interdependence. Outcastes play the drums in festival processions while the Moslems let off the fireworks for Hindu festivals as well as their own. The Outcastes can circumvent the difficulties of life by migrating to the towns where they create yet another sub-caste. Also agriculture can always absorb non-agricultural sub-caste members. Occupational specialization gives each group a vested interest in the village environment as a whole because it is there that it enjoys security in its monopoly.

7.4 States with segmented religious systems

Western countries characteristically recognize Christianity as their basic religion but accept that there are many doctrinal and confessional subdivisions within it. These subdivisions are so intermingled that there are few large blocks of different faiths and it is usually impossible to attribute any particular environment to any religious group. Only their buildings and surroundings are distinct and become significant to the environment when the practitioners are gathered there to carry out religious activities. Their main environmental influence is therefore on their immediate surroundings.

Comparable subdivisions also existed in medieval Europe but were due to administrative, rather than doctrinal divergences. The Church owned enormous areas of land which were only indirectly under its control. It is often seen as a single entity, but each diocese and monastery owned and managed its own land independently. The climate of religious opinion may have influenced the use of these holdings. The Benedictines had different policies from the Cistercians and neighbouring bishops only coincidently had similar land use policies. Today, the ownership of a Roman Catholic parish church is in the hands of the diocesan bishop but he must use this ownership in a way that accords with the views and feelings of his parishioners. Individual establishments could evict tenants and thus change the pattern of land use. The Church of England today owns considerable real estate besides the land on which its churches and schools stand. The Church Commissioners rent some out to other organizations. They have no control over the use of the land except what can be written into the lease as a legally and commercially justifiable restriction. Sometimes a decline in church membership, particularly in city centres, forces the sale of old churches and chapels for conversion to secular uses.

If any religion wishes to occupy a more specific environmental niche in which to practise its way of life in social isolation, the members can rent or buy properties for specific purposes. The enclosed Benedictine nuns of Stanbrook Abbey near Worcester run a specialized printing business. Its large garden has long provided a small particularized environment. Similarly, much smaller Protestant houses of prayer such as Elsie Briggs House near Bristol and the Buddhist monasteries recently started in both England and Scotland provide spiritual support. Once these establishments go beyond family size and take in outsiders or outsider trades, they affect the environment. This brings them under the control of state agencies concerned with safety laws for industrial machinery and food preparation regulations. The former could affect printing presses in monasteries and convents, the latter for houses of prayer taking in weekend guests.

Religious organizations also relocate buildings. The widening suburbs and new towns created in post war Britain all have new churches, as do some cities in which large areas were cleared by bombing. Religious subdivisions have required that Liverpool have two cathedrals, both of which impact the environment by drawing in specialized workers and large amounts of materials, attracting population and rearranging directions of movement.

In mixed religious societies such geographical niches often change rather than disappear. A Baptist chapel that is turned into a private house probably retains many of the same environmental functions as it did when it was a religious meeting place. A change in theological emphasis and socio-economic pressures have turned some central city churches in Britain into social centres for the homeless, bookshops and concert halls and led to tarmacking cemeteries to make playgrounds.

7.5 The environment in religiously divided societies

There are some societies in which religious segmentation also involves hostility between the different groups. This can lead to internal conflicts whose main effect is to speed the segregation of the groups and so retard overall economic development. India is an example. Despite its secular constitution, many areas show an effective geographical separation not only of Moslems from Hindus, but between the many caste groups within the latter. One such subdivision is the Untouchables whose living conditions have benefited from positive discrimination under the constitution.

The Hindu majority in India has often been involved in violent clashes with the large Moslem minority. This violence has a long history. It antedates the colonial period and has even increased after independence with the division of the Indian subcontinent into two states. It has continued throughout the time of the three wars between them and the continuing confrontations in Kashmir. In 1923–27 Banaras had a large Moslem population and there were 117 serious Hindu–Moslem disturbances leaving 450 dead and over 5000 injured (Pandey 1990, 94–134). This was despite the fact that the two communities were economically indispensable to each other and that as individuals there appeared to be no persistent interpersonal tensions. However, in a study made in rural north India half the cases of malignant spirits infesting Hindus were diagnosed as being Moslem (Kakar 1990, 135–45). There has also been persistent violent hostility between Hindus and Buddhists in Sri Lanka. This has led to population movement out of the areas dominated by one or other of these faiths. It has also brought killing, destruction of property and increased the difficulties of farming in the disputed areas.

Severe conflicts have also occurred within Hinduism. Caste Hindus committed 37 000 atrocities, known to the authorities, against the Untouchables between 1960 and 1980 (Bayly 1999, 345–57). These included the burning of Untouchable hamlets and their crops, killing the men and raping their women This seems to have been provoked by the increasing opportunities provided for them near to the towns. It was also due to the granting of new leases with legal land rights. These came at a time of rising land values due to the green revolution of improved grain yields. Other castes opposed all such advances because of their threat to caste status.

The use of the vote and the principles of social justice enshrined in the Indian Constitution made inter-caste tensions more visible. This can be seen in disputes over the use of land and other natural facilities. In Gujarat 68 per cent of Untouchables were denied access to potable water from high-caste Hindu wells (Baxi 1990, 215–39). Untouchable women with their children had to wait outside while the health worker dealt with high-caste women indoors (Social Science and Immunization Research Project 1988, 148).

In recent years Hindus have come to see other religions as more organized so that from 1970 they improved their organization as a reaction to supposed petro-dollar support for Moslems from overseas (Veer 1994). This resulted in larger processions and festivals involving the middle class. In some places Hindus tried to grab Moslem graveyards for cattle grazing, while the religious ghettos of the cities were duplicated in the towns (Banerjee 1990, 45–58).

These internal stresses in India have had significant environmental effects. They have arrested economic development and led to increasing communal segregation. They have worsened India's relations with neighbouring states and diverted resources into armaments.

Such hostility between religious faiths is by no means confined to India. It has been a feature of Ulster life for centuries and in Belfast there has been a de facto rather than *de jure* separation of Roman Catholic from Protestant occupied housing estates. Only the well-to-do can partially insulate themselves from these sectarian environments. No man's lands of waste ground now separate the two communities. Economic development slows, jobs become fewer, and emigration increases, leading to a vicious circle of accelerating decline. The Good Friday Agreement and the temporary re-establishment of self-government somewhat reversed these trends. The division of Cyprus between Christians and

Moslems resulted in the deliberate desecration of Christian cemeteries in the north (Figure 3.1).

The Jewish occupation of Palestine and continuing official and unofficial expansion into Moslem areas continues both active and passive hostility between two societies expressed in religious terms. Disobeying military orders, Jewish squatters have conducted aggressive prayer sessions at the Moslem Tomb of the Patriarchs in Jerusalem and there was an incident when Moslems at prayer were massacred. On the other hand the Arabs under Hashemite rule in Jerusalem systematically desecrated Jewish synagogues, graves and shrines (Sachar 1987, 16, 111). The Jewish environmental assertiveness may be increased by the uncertainties of the Israeli political system which makes coalitions inevitable and increases the power of fundamentalist minorities out of all proportion to their popular votes. A new contentiousness has been noted among English Jews leading to a polarization between liberals and conservatives (Brook 1989, 155–70).

These trends have had conspicuous environmental effects in Israel. Jewish and Arab housing is largely segregated. The former is modern and of Western style. The latter is more traditional and characterized by the greater confinement of women. The associated communal tensions separate economic activities and inhibit their expansion. Even within the Arab community there is some persecution of Christian groups. So many of the latter have emigrated over the past generation that there has been a marked shift in the Moslem/Christian ratio, in turn impacting housing, occupational distribution and land use.

Many contemporary African states are divided between the main 'imported' religions and their local independent offshoots. The latter are further divided between overt regular activities and parallel covert religious beliefs which they invoke mainly for crises. This is unlikely to divide those nations politically because they lack the representative structures capable of sustaining such divisions. However, when a nominally Christian elite has the political power and the rural majority is excluded as in Sierra Leone and Liberia, there is a revolutionary situation but not one normally expressed in religious terms.

7.6 The effect of religion on agriculture

Most of the world's population still depends on agriculture and is also religious. The relationship between the two is both comprehensive and deep. It is especially true in simple rural societies which make environmental choices at an individual and family level. This is because such societies are less dependent on investment and bureaucratic regulations. Religion tends to be an important determinant of what they do with and to the environment. They generally have a strong ethos of working for the common good, but often embedded in an atmosphere of intense inter-village competition frequently expressed in ritual within the cults of their patron saints (Skar 1994, 110; Sallnow 1987, 125–6). Japanese villages gain a sense of community from rituals to persuade the earth god to arrive and depart (Beardsley et al. 1959, 189–93). In pre-Communist China the farmers participating in such communal rituals had to accept their subordination to the dominant landowning elite so that each household's rituals for their kitchen god were at the expense of any community god (Stover 1974, 132).

Such communities generally see their survival and their use of the environment within the context of a spirit world often inhabited by their ancestors. They see it necessary to restrain individualism by such means as enforcing the preservation of spirit groves around springs, burying or abandoning the dead rather than using the wood for their cremation, by restraining the performance of costly material rituals, and maintaining the use of water temples, as in Bali. In such communities it is difficult to find any activity which does not have within it some religious element.

By contrast, Jewish, Christian and Moslem lands see themselves as communities living out God's word. They reject the concept of the intrinsic spirituality of nature and believe in acting on the environment in ways that are determined by the divine will. They accept that each individual is subject to the collectivity in so far as it does not transgress his or her responsibility to the deity (Ellen 1978,196).

But some distinction must be made between Jews and Christians on the one hand and Moslems on the other. It results from the Islamic doctrine of *kismet* (fate) and its human response, fatalism. Muslims tend to tolerate rather than combat unfavourable circumstances. They accept authority with resignation, and often lack a sense of involvement with it. Fatalism plays a large part in their lives, and helps explain the lethargy and lack of progress which, until recently at least, has for centuries characterized Moslem countries (Anderson 1950, 82). God's remoteness and unknowability also lead to a less radical and incisive view of human wrongdoing than in Christianity. Sin is seen as disobedience to an arbitrary will whose motivations are veiled, not an antagonistic rejection of self-sacrificing love. The basic human problem is therefore not defiance and rebellion, but weakness and forgetfulness. This reduces the moral gap between God and humans and decreases their need and incentive to make radical changes in thinking and behaviour.

The effect of this fatalism and more restricted view of sin has been to undermine the innovative spirit and the practical enterprise which should spring from it. Land use practices in Moslem lands tend to be more than usually conservative and conformist. Responsibility is seen as lying outside an individual's control. Quint's study of an Iraqi village illustrates this (1969). Um en-Nahr lies about 60 km south-east of Amara in the rice-growing area. It contained 183 people much afflicted by poverty, disease and ignorance. They desired and believed in 'progress' but saw it as something coming from outside to which they in no way contributed. They did nothing towards it. Decision-making within the village was confined to certain norms. There was little concept of self-help. A respected man (*rajil khayr*) was a conformist par excellence, pious, hospitable, truthful, loyal to his group, entirely predictable and with only a limited range of work he could do without shame (*'ayb*). He did not see enterprise as part of his role. He could not face a situation of status by achievement.

Also, Islamic inheritance law has important effects on landownership and hence on the environment. The individual can only dispose of one-third of his estate, The other two-thirds goes in fixed proportions to his legal heirs, males receiving twice as much as females. Within this there is a difference between Sunni and Shi'ite rules. The former only allows claims from paternal relatives though modified in favour of near female relatives. Shi'ite practice regards paternal and maternal connections as equal. The main environmental effect of these inheritance laws is to subdivide estates. This works well enough for money and animals which can be divided up but verges on the disastrous when applied to farming land, trees and houses as it results in endless subdivisions such as land on which two palm trees stand and shares in a single olive tree. In Upper Egypt a family in the 1920s might have owned a farm of more than 30 acres but by the mid 1950s few families owned as much as 5 (Ammar 1954, 23). An Egyptian dignitary wrote in his memoirs that he owned 60 acres whereas his great-grandfather had owned 600 (Mubarak 1887–9, x, 98).

Land has an emotional appeal so that the common-sense solution of sale and dividing up the money does not appeal. To retain a small plot of the family land is to retain a hold on other family members who have the rights to other parts, a form of social insurance. Through various inheritances the holdings are unlikely to be in one locality, so not only is there time wasted going from one plot to another but land is taken up in making boundaries. Bangladesh adopted the policy of consolidating holdings. Although an economic necessity it clashed with Islamic property laws. It brought into doubt how much an individual would gain from any such action. Also, the advantages would be only temporary since the whole process of subdivision among relatives and widows would start again after the death of the owner of the consolidated farm.

The pattern is somewhat different in small Christian agricultural communities. They have usually developed and retained structures for cohesion which maintained their religious isolation for long periods and also made significant advances in agricultural practice. Among the earliest examples are monasteries which favoured self-sufficiency as a way to put Christian principles into action. Their earliest influence on the environment was in the land immediately around their special buildings. Roman Catholics saw the obligation to work as a divine law and the work of the monks exemplified this in a way that all could see. During the Middle Ages in Europe the Church was virtually the only model of ordered, methodical and systematic government available, so it is not surprising that its influence was impressive.

Medieval monasteries colonized unused land and organized farms larger than neighbouring peasant holdings. The monks sought to give workers on the land protective regulations to increase their efficiency as farmers and also to provide them with some personal security. This led to a release of serfs and a change in their conditions of employment. Such estates developed agricultural science, improved forestry and scientific breeding, created model farms, experimented with new crops, and regenerated agricultural production (Boissonnade 1927, 157).

In Mexico the Church's missionary orders initiated an astonishing range of changes in the environment. They wanted to grow the food to which they were accustomed and which followed the demands of their particular denomination. Roman Catholics needed wine for the celebration of mass and fish on Fridays and during Lent. Thus they grew vines in Tanganyika and created fish ponds in many places. The Augustinians brought fruit trees from Castile to the Americas, in itself a remarkable achievement considering the shortage of fresh water on the long voyage. They were, however, always short staffed. The cultural isolation of the missionary, who was often Spanish, as well as his sense of technological and social superiority, tended to reduce his influence.

In general, missionaries of any denomination saw the social and physical environment into which they had moved as both disorderly and threatening. The binocular vision of Christian and native cultures as demanding some equality of understanding was not always understood in missionary work. They would always have seen their own culture as more orderly with its roads, water supply systems, boundaries to fields and legal titles to everything. Even in late colonial Africa, missionary orders obtained titles to land for building churches in remote areas where no legal titles existed and where they were totally dependent on the goodwill of the people in whose environment they proposed to build.

A Dominican priest taught the Mexicans the systematic cultivation of cochineal. Other Dominicans started the Mixtecs raising sheep and goats, planting mulberry trees, and raising silkworms. By 1542 the diocese of Oaxaca was producing 9000 lb of silk per year until this new industry was undercut by the Chinese.

Such projects were necessarily linked to irrigation. Father Templeque at Otomba near to Mexico City over 17 years constructed an irrigation channel over 30 miles long which is still in use. However, in what is now New Mexico, the missionaries only thought in terms of irrigation in the riverine areas where it could produce crop surpluses. Therefore they had little influence on the Pima, Zuni and Navajo Indians, living as they did in semi-desert environments, to whom such methods were inapplicable. The Roman Catholic Church never managed to convert these desert Indians who largely retained their own religious methods of relating to the shortages of water and rain.

In short, over Latin America as a whole, a small number of unusual and vigorous priests made substantial changes to the environment for the benefit of those for whom they felt spiritual and material responsibility. The immigration of these men, who rarely returned to their own countries, was beneficial to the Indians (Ricard 1966, 142–50).

The Anabaptists in seventeenth century France, precursors of the contemporary American Amish, transformed sterile land into excellent pasturage. They had been forced to develop their own farming methods in previously unproductive regions and climates when they had been disenfranchised politically. In the seventeenth century they were practising the rotation of crops, stable feeding of cattle, the irrigation of meadows, the use of natural fertilizers and the growing of clover and alfalfa as a means of restoring soil fertility (Hostetler 1993, 115). It was only when some Anabaptists migrated to America that they had for the first time the choice of soils and thus benefited from the practices that they had been forced to develop in Europe as ostracized believers. When they were expelled from France, the local administrators in Alsace appealed to the French government against this order as it would mean an economic setback for the region (Seguy 1973,182).

Some religious organizations have significant influence beyond the range of their immediate habitats. The ascetic Cistercian foundations were built in isolated places to avoid too much contact with the laity so that, although they had property in some 40 English and Welsh towns, they had no abbeys in them. They tended to open up previously unworked land and their success in this made them aware of the need for extending existing market facilities. Since they had a highly organized system of isolated community houses and farms worked by a labour force of lay brothers, they were often responsible for the building up of small new towns near by. They helped the authorities to plan and found 44 towns on Cistercian land in the Garonne valley between AD 1252 and 1325 and nearly all of the 24 planned new towns on their landholdings (Beresford 1967, 131–2).

The task of reclaiming marshes in England fell largely on ecclesiastics until royal interest took over at the end of the thirteenth century. Under the Archbishop of Canterbury, they drained about 23 000 acres of Romsey and Walland marshes, and parts of the Somerset Levels and Fenland (Donkin 1973, 104–5). They were involved partly because they had the only available organization with the structures to oversee such work. The process increased the wealth of their religious institutions.

Sometimes, monastic communities used their lands to develop their own rules for spiritual life and to ensure that they could live as far as possible independent of charity. Such charity must have been very sparse especially for isolated Cistercian houses. Then as now many religious organizations were unable to survive on occasional donations and needed 'unearned' incomes from charitable bequests of money and property in order to survive as well as implement their Christian obligations to others.

Monasticism has, however, indirectly affected the environment through the discoveries of its adherents. Monasteries pioneered improvements in animal husbandry. Gregor Mendel, a Roman Catholic monk, pioneered the genetics which form the basis of the animal breeding universally used in modern agriculture. Adam, a contemporary Benedictine monk, is a world authority on the development of productive strains of honey bees.

Some faiths have unusual views on the agricultural environment. The Jains abstract themselves from any direct involvement in it except for their temples and religious households. They avoid destroying trees because they believe they feel pain when sawn. Similarly, the Buddhists' reverence for life causes them to avoid killing any creature. U Nu, the

Prime Minister of Myanmar, describes being unable to flush the lavatory because there were ants in the bowl.

Such impractical themes would seem to prevent these religions achieving adjustments to their environments when their followers have to survive as subsistence farmers. This leads inevitably to 'popular' religious practices which retain some elements of these faiths but take notice of their practical needs. Only a very small minority will put into practice extreme beliefs which go against their social and indeed biological survival. In practice most religious groups accept a framework of spiritual injunctions but do not follow them so closely as unreasonably to interfere with the rhythms of agricultural life.

Both Amish and Hutterites illustrate this. They have communities based on biblical precepts which have been significantly successful in retaining their members over several generations. They have made a religious imprint on the environment in which they live, while enjoying a different lifestyle from their neighbours who do not share their beliefs. The Amish, and probably also the Hutterites, in using lower levels of technology and disdaining the attractions of consumerism, have consistently managed to achieve higher standards of living through good farming methods than their neighbours who use more advanced technology and accept consumerism (Bennett 1976, 277–81). The fact that the leaders in such communities feel that their status is assured and their lifestyle controlled by their religion gives them an overall confidence in dealing with their environment. Mixed farming and fallow periods combined with communal help, advice and support of the community for those starting out on their own, give them an environmental advantage.

7.7 Settlement patterns

Some religions have gone beyond merely creating institutions with landholdings and introducing new agricultural methods with opportunities for any who cared to listen and learn. The Roman Catholic Church in Mexico, for instance, created villages for its followers. But the leadership always feared Indian uprisings. The roof of the cathedral in Merida in Yucatan has embrasures for cannon.

The Jesuit communities or 'reductions' in Paraguay, Brazil and Argentina differed from monastic establishments and Amish settlements. They did not reflect the needs of the few Jesuits involved who by the rules of their order were not communally organized. The missionaries saw the way in which the Guarani Indians and others were being both persecuted and taken into slavery in large numbers. They concluded that the only way to affirm the dignity of these people and give them a chance to lead Christian lives under reasonable conditions would be to form communities to separate them from the encroaching whites. Thus during the period from 1607 to 1767 when they were expelled from both Spain and her colonies, a small number of Jesuit priests and brothers, probably not much more than 100 at any one time, eventually developed and maintained 57 settlements or townships in which some 50 000 Guarani Indians came to live (Caraman 1975, 314).

Each reduction constituted a self-contained unit. Two Jesuits took charge of the settlement where they served as priests, doctors, teachers and work overseers. Part of the land on which the settlement stood belonged to individual Indians, but most of it was the property of the community. On their own land the Indians could plant what they wished, but the priests supervised the cultivation of the communal 'God's field'. The harvest provided the resources for maintaining the sick, aged and infirm as well as providing funds to build houses, churches and farm buildings and pay taxes. The missionaries themselves did not share in this distribution since they lived on small stipends provided by the Spanish king (Fulop-Miller 1930, 290–2).

Each settlement centred on a plaza with a large stone baroque church. It had an elected Indian leadership and the Jesuits trained Indians in a number of sophisticated and technical skills. Only the Spanish governor and the bishop had access and visiting merchants were kept residentially segregated. After several attacks by slave traders, the Indians were armed and trained to fight and were successful in several skirmishes. Once the Jesuit supervision was removed these settlements gradually declined and their populations dispersed so that all that is left today are some striking ruins such as those at Trinidad in Paraguay, San Miguel in Brazil and Santa Ana in Argentina (McNaspy 1982). The extent of these ruins and the fact that they were built out of their agricultural profits testify to the viability of the settlements.

When the Jesuits were expelled, the Spanish authorities carried out an inventory of the property owned by these 57 settlements and recorded astonishing totals such as 698 353 cows, 44 183 oxen, 240 027 sheep, 28 204 horses and 45 646 mares (Brabo 1872, 668–9), testifying to their success for over a century and a half. They had had a significant influence on the environment in turning large areas into settled agriculture for individual and communal farming. The whole inspiration for this was based on the religious ideal that the Indians were capable of becoming Christians and developing a lifestyle independent of colonial oppression. These particular Roman Catholic influences on the environment in northern Brazil continued up to the 1980s.

In 1680–82 the Portuguese authorities placed the indigenous peoples of Brazil in labour pools administered by European missionaries until this was stopped in 1755. But later exploitation of Indians and atrocities against them led to the founding of the Salesian mission in 1914 to offer them sanctuary and protection from forced recruitment. Mission settlements once again became the only refuge for the Indian so that by 1978–81 some 10 000 were living in or associated with three mission villages. These were originally established to provide labour for extractive industry but persisted long after it had proved unprofitable. They were allowed to continue to occupy and fortify the potentially disputed frontier zone with Venezuela. The missions served the national government by organizing the Indians into productive units, instilling some sense of Brazilian identity, and cultivating land where they provided a permanent Brazilian presence (Chernela 1998).

The Jewish settlement patterns in Israel provide a contrast in being a permanent colonization by independent individuals with religious motivation. Their aim has been to establish a society in which Jews would provide the whole basis of their life rather than be dependent on others. This came in part from their previous experience of being confined into ghettos or restricted to particular occupations. There are four distinct patterns of agricultural enterprise in Israel. The extremes are the *moshava* and the *kibbutz*. The moshava is a village of European type with land, buildings and farming installations all privately owned. The kibbutz is a communal settlement in which all members live in close proximity and work the land on a collaborative basis. The kibbutz markets the produce, pays wages and arranges housing, providing communal washing and dining facilities, libraries and a children's school. All the kibbutz members meet to make important decisions. In between these two extremes are the *moshav* and the *moshav shitufi*. The moshav is a smallholders' settlement where each member has his own plot, but is constrained in a number of ways. He cannot sell the land without the permission of the moshav committee and of the national institution which owns the land. He must work the farm himself, help neighbours when necessary and market produce jointly with other members. The moshav shitufi is a cooperative smallholders' village where production is carried out communally as in the kibbutz, but where family life is self-contained, as in the moshav.

The environmental consequences of these four different types of settlements become clear in aerial views. In general the moshavot show dispersed settlements of single family farms, kibbutzim larger fields and closely nucleated settlements with all buildings together. The moshav and moshav shitufi pattern is of a central area of housing with private plots radiating outwards from them (Orni and Efrat 1964, 213). All the settlements on the national frontier of whatever type have a defensive function.

These Jewish settlement types represent spatially and on a small scale the universal divergence between the communal and individualistic approaches to life. The acceptance of central authority and the strong spirit of cooperation derive directly from the basis of Jewish faith in the Torah. The differences between the different types of settlement reflect the extent to which the need for communal cooperation is seen as outweighing the preservation of family independence.

But Jewish agricultural settlements did not come to an unoccupied environment. The Arabs had occupied Palestine for many centuries with land use patterns with a much higher emphasis on nomadic pastoralism. A satellite view shows a strong contrast between the intensively farmed Jewish areas and the more extensive farming and unrestricted grazing of Arab areas. There is also a marked contrast in standards of living between the two groups. Their different approaches are in part due to religion. The resulting competition for land has contributed to long-term insecurities which have led to conflicts which divert resources from agriculture into defence.

8 Religious Influence and Population Pressures

8.1 Factors connecting religion and population levels

Religious factors affect the occurrence and toleration of high population densities. This may be temporarily acceptable at focal sites like Mecca or Lourdes but not in people's home areas. The density of population in a monastery may be high but there are institutional arrangements which control the resulting stress. The same density in a newly established commune based on ethical if not religious principles would find the situation more difficult to manage. A Hindu joint household with sons and their wives living with their parents is not a particularly stressful situation for those involved, whereas it might not work for English Christians.

Densities depend on the economic costs of living space but also on birth rate, death rate and migration. These are all influenced by religion. Moral rulings can affect birth control by restricting intercourse to monogamous relationships within which they encourage it at certain times and discourage it at others. Birth rates are high in poor areas where children are an economic investment and there are religious restrictions on contraception and abortion. In areas with high child mortality and an absence of efficient welfare institutions the poor will always produce as many children as they can in the hope that enough will survive to sustain the family future. Birth rates are low in areas where the raising of children involves high costs and where methods of birth control are widely available. Polygyny is condemned by most faiths, but where tolerated it can lead to larger families but a lower number of births per woman. A culture supporting large families also encourages population growth. Death rates depend on lifestyle which in turn reflects religion. In general, rational faiths which inculcate balanced diet, exercise, rest and religious belief correlate with longevity.

Pilgrimages can be dangerous because of the localized high densities that they bring. In 1893, 33 000 pilgrims died in and around Mecca from cholera and the dead remained unburied. In 1987 400 died in violence initiated by Shi'ite terrorists and in 1990 and 1994, 1426 and 252 pilgrims lost their lives in stampedes respectively.

Religion-induced migration likewise has an important effect on population densities. It can be near and temporary or far and permanent. The former includes attendance at mosques and churches or visits to shrines and other sacred sites. The latter involves migration to religious states such as Israel. This may not work out as hoped so that the would-be migrant returns to the original country. The Israeli government organized transportation of the Jewish Falashas from Ethiopia to Israel. This can be presupposed to be permanent since, apart from the religious motive, their standard of living will have been raised by the migration. Another example is the vast exchange of populations between Hindu India and Moslem Pakistan at the time of partition in 1947.

Motivations for religious movement are part personal and part social. The reason for going on the Hajj might be that the pilgrim's father-in-law had provided the money and expected him to go. The lonely widow may join a group from the local church going to Lourdes. Whatever the motive, the environmental results are the same.

8.2 Psychological factors in religion and population pressures

Although it is sometimes possible to assess the relation of religious practices to population pressures, it is more difficult to assess the psychosomatic factors which underlie the performance of such practices. Religious conformity can induce a feeling of well-being, that a person is in harmony with their religious universe. It gives them a 'feel good' factor. This encourages people to carry on and maintain population levels which they might not be able to do if their view was more fatalistic. They instinctively follow a divine injunction to 'be fruitful and multiply'. A study of the parents of dead children found that the Roman Catholics who believed that it was God's will survived better biologically than Protestants who blamed themselves for failing to prevent it. Moslems have a stronger form of fatalism. A study in Saudi Arabia showed that people accepted that malignant disease was believed to be beyond the physician's powers (Bahakim 1987, 22–6; Bedikian and Thompson 1985, 16–17; Panter-Brick 1991, 1295–302). The Ik of Uganda may come near to this with their apparent total communal and individual passivity in the face of adversity (Turnbull 1994).

The Ottoman empire was notorious for its fatalism when confronted by the plague. In the sixteenth century it made no effort to take precautions, to isolate victims or to prevent the transmission of infection (Slack 1985, 49), even though with medieval medical knowledge they could not have done anything effective anyway. The Turks stayed where they were and died, while at least the Christians moved away from the centres of infection. The Moslems had become acculturated to the phrase 'in sha' Allah' (if God wills it) and thus were pushed towards inactivity, while the Christians were moved to greater activism in seeking preventive measures.

Although this feeling of impotence may have a stronghold in contemporary Islamic societies, Christian thinking has often followed Job's comforters in seeing disease as punishment for sin, and moral reformation as the key to recovery. The Black Death was widely regarded as a visitation by God because of the general sinfulness of society (Ziegler 1970, 36 and 39). In medieval times it was believed that personal self-abnegation and penance could reverse this but the processions of people whipping themselves in atonement for their sins certainly spread the disease until they were stopped by Papal edict. It was only with the acceptance of post-Reformation rationalism that a scientific approach to understanding and controlling disease was adopted.

Mahatma Gandhi in 1934 shared the general belief that calamities come to mankind as a chastisement for their sins. The Mormons regarded the cholera pandemic of 1832–33 as a divine chastisement for wickedness (Bush 1986, 401) and it is a view among some Christians that AIDS is a divine judgement on sexual sin.

The powerful Aladura Independent Church in West Africa also holds to the conviction that illness is related to sin and that the use of medicine is a form of self-deception (Peel 1968, 129–35). This assumed connection between sin and disease may well be sufficiently widespread to have an influence on public health and population levels. The head of the Aladura Church during a smallpox outbreak stated that sanctified water was the best protection (Turner 1967, 142).

The Christian Scientists view creation as purely spiritual and therefore that healing must recognize this. They believe that disease is in the mind (Eddy 1906) and can be controlled there. This may have had some demographic effect since it has been estimated that their death rate from malignancy was 6 per cent higher and at a much younger age than would have happened if they had allowed themselves surgery (Wilson 1956, 43–60). Other groups such as the Jehovah's Witnesses, who refused blood transfusions until recently, are too small in numbers to have a significant environmental effect.

The Mormons and Seventh-day Adventists have a faith and lifestyle which give them an improved health record over other groups and faiths. It also encourages their psychosomatic stability. Although they are relatively small groups, both have an influence which raises the health and longevity and thence the general prosperity of their home areas. This is visible in Utah and parts of southern California respectively.

Change has been more rapid in the last generation than ever before and has exacerbated the psychosomatic problems in adapting. There are said to be five possible ways of coping with the problems of change: conformity, ritualism, retreatism, rebellion and innovation (Merton 1964, Chapters 4 and 5). All have a religious aspect and affect the environment. The first three accept the changes without protest and adapt to them in different ways. The last two contest them and seek to change the environment, as can be seen in the recent 'environmental' and 'green' political campaigns. These may at times be at the expense of neighbours who do not share these beliefs.

8.3 Inter- and intra-society religious violence

It is an unfortunate fact that religious differences have underlain most human conflicts. Many have been extremely destructive of human populations, both in absolute numbers killed and in their proportion of whole populations. This is a reflection of the central importance of religion in the way individuals relate to the cosmos and the resulting tenacity with which they struggle to establish and maintain what they see as truth. The Taiping Rebellion in China in 1851–64 under a leader claiming to be the brother of Jesus Christ (Boardman 1952) may have killed as many as 20 million people so that in some cities only dogs survived (Wakeman 1966, 3).

Tamurlane (born AD 1336) is said to have killed from 'piety' (Grousset 1970) about 80 per cent of the population along the northern belt of Arab–Iranian societies in central Asia. His men are reported to have piled 80,000 heads outside the walls of Isfahan (Lamb 1932). The area has even now not fully recovered from this and the sedentary population has only inadequately developed the environment. At the Christian siege of Damietta in 1219 it is said only 3000 Moslems out of a population of 80 000 survived (Oliver 1894). These were events that significantly reduced populations.

The consequences of religious wars in the Near and Middle East as well as in central Asia are magnified by the severity of the environment. These areas have irregular rainfall both in overall amount and in seasonal distribution. About one year in five it is too dry to grow crops and people suffer severely. Even a good annual total of rainfall can be a disaster if it does not come at a critical time for crop growth. It is most needed in October–November to establish the cereal crop and in March–April to bring it to harvest. At other times much runs uselessly to waste. When coinciding with drought years military destruction can have disastrous demographic consequences. Tunisia, Israel, Jordan and Syria were once granary states in which persistent warfare, largely about religious questions, reduced their capacity to organize communal support for agriculture. Even the Prophet Mohammed remarked on the remains of past civilizations in Arabia (e.g. Koran 27: 42). Near Qal'at Sim'an in Syria there are the ruins of 42 ancient towns in comparison to the 14 hamlets there which are still occupied. Religiously motivated violence in the Islamic period reduced populations throughout the whole belt from Morocco to Turkestan, severely hindering technical and social progress. Unlike relatively well-watered Europe, the arid climate of the area retarded recovery (Russell and Russell 1968, 10). Where agriculture depends on irrigation, the risks are greater. This requires a complex technical system highly vulnerable to destruction during wars. When such destruction occurs whole civilizations can fall, as seen in the histories of Egypt and Mesopotamia.

Within Islam hostility between Sunni and Shi'ite has a long history in which the latter have been regularly persecuted. The recent war between Sunni-led Iraq and Shi'ite-led Iran had religion as a contributory cause. Despite the enormous loss of young men's lives the environmental effects were unusually low as most of the combat took place in the desert. But Shi'ite groups in India, particularly the Isma'ili sect led by the Aga Khan, have avoided tensions by keeping themselves socially and economically distinct. The Ahmaddiya are a modern Islamic sect, claiming particular insights from a nineteenth-century prophet, who reject the use of violence in the jihad. They have been persecuted in Pakistan as anti-Islamic.

Mohammed united political and religious authority and this has sanctified their union in Islamic society ever since. It has made for homogeneous societies but has meant that internal differences in all branches of Islam have had both dimensions. Political differences have deeply involved religion; religious differences have deeply involved politics. It has tended to exacerbate conflicts and make them more destructive. Since Sunnis are in a majority within the faith their campaigns provide most examples, notably the Mahdiyya in the Sudan (1883–98), the Wahhabi rebellion in Arabia under Abdul Aziz Ibn Saud (1919–24), and the Taliban takeover of Afghanistan in the 1990s.

The Mahdiyya saw a great reduction in the Sudan's population. Mohammed Ahmed Ibn es-Said Abdulla, born 1844, developed a mystical interpretation of Islam and created a vast Islamic state extending from the Red Sea to central Africa He had a great appeal to the millions of Islam by opposing infidel influences and seeking to return the Sudan to a purer faith. He captured Khartoum in 1885. After his death his chief lieutenant the Khalifa Abdulla Ibn Mohammed assumed his mantle and ruled with such ruthlessness that the population was reduced by more than half before General Kitchener's final reconquest of the country in 1899. It required a generation under the administration of an external condominium of Britain and Egypt for the country to recover from the economic devastation thus caused.

The Turkish killing of a third of the Armenian population of Asia Minor in 1915–17 (Boyajian 1972, 103) was environmentally destructive and was further exacerbated by emigration. In the Lebanon, there are continuations of circulating tensions between Maronite Christians, Sunni and Shi'ite Moslems, and the Druzes in which their differential birth rates are an important precipitating factor.

Pogroms in Russia and Poland from the end of the nineteenth century killed many Jews, rising to an estimated 60 000 in 1917–21. The main culprits were the White Ukrainian Army in 887 major and 349 minor attacks (*Encyclopaedia Judaica* 1971, 13, 694–701). Finally, the Holocaust killed an estimated $6\frac{1}{2}$ million Jews between 1939 and 1945. This not only destroyed the urban locations in which their communities had lived for centuries, but also damaged the rural areas which served them.

The fact that only 10 per cent of all wars between 1820 and 1952 had neither a Christian nor a Moslem component (Wilkinson 1980, 87–92; Richardson 1950, 237–55; Rupesinghe 1994, 21) has led some commentators (Wilkinson 1980, 87–92; Wright 1964, 179–80) to conclude that both Christianity and Islam were naturally aggressive. These conflicts occasionally brought these two faiths into conflict as in Lebanon and Bosnia, but were equally frequent within each. The two world wars were mainly between Christians; the Mahdiyya in the Sudan and the war between Iraq and Iran were mainly between Moslems. But the points in contention have generally been over land and jurisdiction rather than purely religious issues.

Religious distinctions can be expressed in dress and lifestyle, so that one group of people will treat another as outsiders even when they are in fact neighbours. This can enhance feelings of personal loyalty within groups and exacerbate their differences. For example, if violence had occurred within either the Christian or Moslem communities in Cyprus, its perpetrators would have been regarded as psychopaths, whereas in the inter-communal fighting they were readily identifiable as defenders of their way of life and communities (Loizos 1988, 634–53).

Eastern religions have been less concerned with promoting an exclusive confessional viewpoint. Hinduism, until it was confronted with Islam, had a heterogeneous population of believers with a multiplicity of gods. But even in that type of religious environment intra-religious riots can occur. They can even be initiated by incidents as trivial and localized as the theft of a Hindu statue by other Hindus in the same village. Most recently the faith of the Indian Hindu has become simplified, politicized, mobilized and ascribed a transcendent value above other faiths, and even of the secular laws of India. This transformation has brought a continued rise in religiously motivated violence there (Brass 1997, 260–2).

Christianity has probably had the most environmentally destructive internal wars. Between 1820 and 1950, 90 per cent of all deadly quarrels within a particular faith were between Christians and 76 per cent between religions which had Christians as one of the combatants as opposed to 35 per cent with a Moslem combatant. The percentages are even higher before 1820 with parts of Europe periodically devastated by wars between Christians. The main exceptions to this are the Crusades and the intermittent 200-year reconquest of Spain from the Moors. This is less true today perhaps more because of a decline in religious observance in western Europe than to ecumenical negotiations.

Political, social and territorial changes follow from religious conflicts. The Islamic occupation of southern Spain was gradually pushed back between the recapture of Toledo in AD 1085 and that of Granada in AD 1492. The resulting dislocation was modified by the slowness of the change and by conversions to Christianity. From AD 1499 the Moors were given the alternative of conversion or exile, although the ecclesiastical authorities in Rome were doubtful of the validity of such conversions since only a few were voluntary. The towns and irrigated lands were predominantly Christian, while the ex-Moslems, called

Moriscos, tended to live in the poorer land in the hills. They were provoked to rebel in Granada in AD 1568–70 with the result that they were rounded up and moved to Castile. In return, Castilians moved into their deserted villages though they failed to develop the area successfully.

The Moriscos, after living nearly three centuries cheek by jowl with Christians, still retained their traditional distinctiveness. In Spain they had grown up alongside a rural and urban population of fanatical Old Christians who have been compared to the poor whites in the southern states of America (Braudel 1973, 2, 780–6). Even if they had become dubious Christians or ex-Moslems they remained unassimilable. They retained their own dress, language, cloistered houses and public baths (Braudel 1973, 2). The ultimate solution was the expulsion to Morocco of the entire Morisco population out of a total Spanish population of about 8 million. This happened in AD 1609–14 after prolonged persecution. Their strongest defenders were in fact the Christian feudal landlords in much the same way as the southern plantation owners in the United States protected their slaves. It left many villages virtually empty, according to a contemporary Spaniard who was mapping the kingdom (Asso 1790, 338). The result was comparable to the mutual evictions of Moslems and Christians from Greece and Asia Minor in the 1920s. It emptied villages and replaced experienced communities adjusted to their environments with a parasitic influx of civil servants, priests and townspeople aiming at colonization by exploitation rather than by settling. Spain lost a hard-working population of agricultural workers and artisans.

This whole episode in which Christian Spain deprived itself of valuable people was provoked by Christian intransigence both at the street level and in the highest echelons of the Roman Catholic Church. It was probably carried out against the environmental and economic needs of the Spanish state. It was not inspired by racial hatred which seems to have been totally absent from the conflict, but by religious and cultural enmity. One can suppose that Morocco gained in the long run from a massive immigration of hard-working competent farmers. This Spanish experience of prolonged hostile contact with another culture within their boundaries formed the background to their attitude to, and exploitation of, the American Indians.

The significance of contemporary religious warfare on population levels has not always been recognized because the death totals are not high enough to merit mass media attention. Nevertheless, they can be highly significant in small communities. In Ulster over 3000 people died between 1969 and 2000 out of a population of about 1.5 million from politico-religious violence. If this figure was applied proportionately to Britain and the USA the totals would have been 120 000 and 520 000 respectively, with that for injured being several times higher (Hickey 1984, 119). This violence also leads to the segregation of the community. In Belfast 70 per cent of the population live in religious ghettos of their own choosing after decades of harassment. The economic losses have been as damaging as in some conventional wars. In the period from 1971 to 1982, Northern Ireland saw 7200 bomb explosions, 3200 bombs defused, 28 500 shooting incidents, as well as over 100 churches destroyed or badly damaged (O'Malley 1983, 10–11).

Religious terrorism can involve the suicidal commitment of the perpetrator. Members of the Palestinian Hamas, for instance, believe that if they give their lives on God's behalf they gain immortality as a result (Rapoport 1982, 4–5). But their suicide bombings in Israel have little direct effect on the environment as they do little to change government action except further to segregate the Jews from the Arabs.

If Communism is treated as a religion, then the total of deaths from its conflicts would be large. In Java in 1965–66 there was something approaching a Moslem 'jihad' or religious crusade against Communists resulting in perhaps 250 000 deaths (Cribb 1991, 8). In Bali there were parallel killings of 20 000–60 000 Communists by the island's Hindu majority (Cribb et al. 1991, 246). It was recorded that some of those to be killed, lined up in traditional funeral clothes in preparation for their own deaths.

Such conflicts involve clergy and other religious leaders. There are newspaper photographs of weapon training in mosques and mullahs carrying Kalashnikovs just as there are pictures of Christian priests shooting in the Lebanese civil war with pictures of the Virgin Mary stuck on their rifles.

Both the Bible and the Koran contain many references to the use of force and include some violent language. Both permit it when the cause is judicial. The battle against and victory over evil is the common theme. While both the Old Testament and the Koran enjoin military action in defence of religion, the New Testament does not, although it frequently uses military terms in a figurative sense. This sort of usage can also be seen in Christian hymns such as 'Onward Christian Soldiers' and 'Fight the Good Fight'. The structures and uniforms of the Salvation Army and Church Army echo this. It has sometimes been perverted for use in religious nationalism (Juergensmayer 1993).

All religious conflicts are environmentally damaging. They cause many casualties and deaths. They lay waste the areas over which the conflicts rage. They arrest development by withdrawing young men from productive occupations and leave a legacy of damaged buildings, military earthworks, war cemeteries and neglected public works.

9
Religious Acts with Consequences on Communal Lifestyle

9.1 Reducing morbidity and mortality

Health teachings are closely connected to religion and are significant to the environment when they receive mass observance.

The Torah is probably the earliest text on health. It contains many injunctions about diet, hygiene and lifestyle. The Jews observed them during their wanderings and there 'was not one feeble (person) among their tribes' (Psalm 105: 37). It is revealing to compare the common-sense rationality of the Mosaic injunctions with an Egyptian medical text of the same period. The Papyrus Ebers recommends several hundred remedies for diseases. The drugs include:

> lizards' blood, swine's teeth, putrid meat, stinking fat, moisture from pigs' ears, goose grease, asses' hooves, animal fats from various sources, excreta from animals including human beings, donkeys, antelopes, dogs, cats and even flies.
>
> To embedded splinters they applied 'worms' blood and asses' dung'. Since dung is loaded with tetanus spores, it is little wonder that lockjaw took a heavy toll of splinter cases.
>
> (McMillen 1971, 11)

The Jews have more or less maintained good health and longevity to the present day through the observance of the Mosaic code. Islam has also largely followed this. The Koran adds nothing new except to emphasize cleanliness (Sura 2: 222), the value of fasting (Sura 2: 183–96) and abstinence from strong drink, carrion, blood and swine's flesh (Sura 5: 3). But other condemned foods, such as the camel and shellfish are sometimes eaten. Jewish principles were also carried over into Christianity, until in early centuries the increasingly Gentile character of the Church caused a reversion to earlier habits. It seems likely that observance of these principles has contributed to health and longevity wherever they are practised, and that societies which have neglected them have suffered.

The scientific understanding of the bacterial and parasitical causes of infectious disease is modern and owes much to the disciplined research approach associated especially with post-Reformation Christianity. It has enhanced the understanding of the effects of some practices deriving from religious belief. For instance, is now known that pork is an unhealthy meat both because of its high fat content and as a bearer of parasites. This seems to be partly due to its lifestyle. The pig is a scavenger and differs from animals of the antelope and ruminant tribes which, although they can live in equally unsanitary conditions, are nevertheless grazers and not scavengers.

Hygiene has a religious basis and affects morbidity and mortality rates. The Torah contains detailed instructions for maintaining domestic and camp cleanliness and dealing with 'leprosy'. This English word, used often in the King James Version of the Bible, probably refers to a range of degenerative bodily and skin conditions. Leprosy was morally associated with sins such as the shedding of blood, taking oaths in vain, incest, arrogance, robbery, and could be countered by using sacred objects for which segregation was enforced (*Encyclopedia Judaica*). Moslems reflect health teachings in small ways such as reserving the right hand for greeting and eating and use the left for acts of personal hygiene and sexual activity, a distinction especially important in areas with limited washing facilities.

Orthodox Jewish limitations on marital sexual intercourse may reduce morbidity and mortality. The low rate of cervical cancer in Jewish women correlates both with the requirement for sexual abstinence for fourteen days over the menstrual period (Weiner et al. 1951, 418–22) and also with male circumcision (Wynder et al. 1954, 1016–25).

The dominating importance of cleanliness is shown by the Mormons who, with no such restrictions, have a cervical cancer rate 26 per cent below the US national average and 50 per cent below the Utah rate for non-Mormons. The data on circumcision, however, are equivocal since the uncircumcised Amish also have low rates of cervical cancer (Cross et al. 1968, 102–8) and the data from India are similarly confusing (Rewell 1957, 821–6). Now it is accepted that the human papillonia virus is involved. Vulnerability to this is affected by the age of first intercourse, the number of sexual partners, and their state of hygiene.

There appear to be a variety of factors controlling blood pressure of which religion is one. Those who practise their faith devoutly have low blood pressure (Larson et al. 1989, 265–78), but sitting quietly and reading a pleasantly entertaining novel lowers blood pressure too. The obvious causes are biological factors such as posture, breathing and temperature. But also important are the psychodynamics of faith and ritual, the social support which comes from being a regular member of a worshipping community and following its specific health-promoting regime (Levin and Vanderpool 1989, 69–78). Mormons, Seventh-day Adventists, Amish and Hutterites all follow such regimes. Their low rates of lung cancer, for instance, result from their abstinence from smoking (Troyer 1988, 1007–17).

Observant Mormons in the United States have a life expectancy seven years over the national average (Enstrom 1978, 1943–51). Such a figure is environmentally significant in Utah where 70 per cent of the population are Mormons and there is a religious system well organized for the support of the old. It is less significant in California where they are dispersed among a much large population of non-Mormons. But nevertheless their better than average health profile shows up even when they live outside the Mormon-dominated Utah environment (McEvoy et al. 1981, 1350–7).

Seventh-day Adventists likewise have an extended life expectancy and are considerably healthier than the members of other religious groups (Wynder et al. 1959, 1016–52; West et al. 1980; Phillips 1975, 3513–22; Fraser 1988, 833–6; Fraser et al. 1991). This is due to a combination of factors; a strong religious faith, a positive emphasis on healthy living, low levels of dietary fat and cholesterol, a well-observed weekly rest day, and an emphasis on family solidarity. It is the totality of these factors which is significant environmentally. The effect of each one in isolation would be difficult to prove since Adventists are generally dispersed in the population and not all observe the dietary principles fully.

Public health has on occasions been related to religious beliefs as a stimulus to public action. In New York, Dr J.H. Griscom, a Quaker and principal health officer, held that hygiene and physiology had a spiritual as well as material content. He drew habitually on arguments from design, taking the view that New York's mortality rate could not be a normal part of God's world because it was so high (Griscom 1857). He campaigned for the improvement of the city's health over two decades, consistently underlining the religious origin of his concern. Robert Hartley, a Presbyterian merchant with the evangelical enthusiasm of youth, was considered to be even more influential than Griscom in initiating social welfare programmes. When attending a lecture on sanitation, he wrote of a desire to dwell on spiritual things (Hartley 1882, 288). The religious convictions of these two men had a major effect on the health of New York City in the nineteenth century (Rosen 1958, *passim*).

It is claimed that miraculous cures occur at religious shrines and were frequent in the Middle Ages. Nevertheless, even if true they were too infrequent to have a significant effect on the environment. The effects of prayer have often been discussed and vindicated in human experience but they are hard to handle statistically because the factors involved are indefinable (Byrd 1988, 826–9; Collipp 1969, 201–4). There are some data that seem to support the validity of charismatic healing but nothing to relate this to the environment (Harrell 1975). Lourdes pilgrims have generally concluded that there has been no physical change in their conditions but 'they are cured of despair, of sadness, of their inability to accept mortality. What is one definition of health? Optimum adaptation to one's environment. In this sense every pilgrim, bodily sick or not, who goes to Lourdes receives a benison of health' (Garner 1974, 1254–64).

A high proportion of epidemiological studies, in which religious behaviour has been one of the factors assessed, has shown that frequent and regular religious attendance is conducive to better health. It seems that direct involvement is the main protective factor. This includes a wide range of activities such as Bible study, voluntary work, holding office in church groups, going to religious schools and of course actually attending religious ceremonies (Levin 1996, 849–54). Spirituality is the underlying factor in all but cannot be measured by the tools of social science. There is evidence that religious involvement benefits middle-class suburban Americans (Crawford et al. 1989, 16–22; Ellison et al. 1989, 100–23) who, for instance, supplement their attendance at church or synagogue with different forms of prayer groups to support their religion-directed activities and help with more personal difficulties (McGuire 1985, *passim*; Wuthnow 1977, 87–99).

9.2 Increasing morbidity and mortality

Religious practices which increase morbidity and mortality are easier to measure than those which decrease them. This is because they are related to shorter-term behaviour patterns. While regular religious practices benefit health, infrequent religious behaviour, often crisis related, seems to be associated with an elevated risk of morbidity (Levin 1996, 849–54). Last-minute conversions appear to have no measurable effect on an individual's chances of survival. As we have seen, pilgrimages can spread diseases and attract opposition from the inhabitants of the territories through which they pass.

The use of water for religious rituals can have benign effects. Judaic washing before meals was associated with lower rates of infant mortality in the nineteenth century in New York. But hand dipping into holy water such as at the entrances to Roman Catholic churches can be a source of infection, even if the water was not polluted at its source (Losito 1946, 463–65). Although Moslems wash their hands, feet and faces before praying, this is often from polluted water sources (Underwood and Underwood 1980, 294; Olivier and Ansari 1967, *passim*), and even makes prayers risky in some places (Farooq and Mallam 1966, 377–87).

Other religions which regularly use ritual water have considered Christianity as unclean and as a step back from Jewish and even from traditional standards. When Jerome came to Rome as an Illyrian immigrant he found a city with 9800 public baths. But he felt that these resorts of profane pleasure would be better converted into sacred baptisteries. The man who had bathed in Christ, in his opinion, needed no second bath. This undervaluing of washing and tolerance of filth has continued to our own day and has reflected discredit on Christianity in the eyes of Jews. Poor hygiene has often accompanied the expansion of any religion into tropical countries with dense populations (Darlington 1969, *passim*).

Religious groups generally prefer their members to marry within the faith. This has little effect on health within the larger confessions. But it poses some dangers within smaller isolated groups by forming closed genetic pools. Both the Amish and the Hutterites almost always marry within their own communities and have unusually high rates of breast cancer and juvenile leukaemia (Troyer 1988, 1007–17). The Amish in particular have high rates of some forms of abnormality such as the Ellis–van Crefeld form of dwarfism. This one community contains the same number of cases as the whole of the rest of the world (McKusick 1973, 1276–84). Also, no less than 12 genetic disorders have first been identified among the Amish. These small religious sects with high birth rates may therefore be held back from expanding environmentally by a load of abnormal members for whom they take personal care rather than shifting them to institutions.

Suicide for religious reasons catches the attention but has little effect on the environment because the numbers are too small except where an entire community annihilates itself. Large numbers of Japanese soldiers and civilians in the Second World War (Toland 1971), including some 4000 kamikaze pilots (Inoguchi and Nakajima 1960) committed suicide. In 1978, 900 believers at the People's Temple in Guyana killed themselves (Hall and Galanter 1989). Suttee, the voluntary self-immolation of Hindu widows on the funeral pyres of their husbands, although forbidden by the former colonial and present Indian governments, was religiously approved. However, cases were never numerous enough, except in upper caste families (Altekar 1956), to affect the environment through population size since even ordinary widows rarely remarried.

Only Judaism allows self-destruction 'kidush ha- shem', the sanctification of the divine name when an individual is faced with an unavoidable threat to their religious principles, even providing preparatory prayers. This accounts in part for the mass suicide of the Masada Jewish community in AD 73 and that in York in AD 1190.

Some devoted religious behaviour may accelerate the local death rate. Eusebius records the extreme self-abnegation of some Christians during the plague in Alexandria in AD 263 when:

> Most of our brethren were unsparing in their exceeding love and brotherly kindness. They held fast to each other and visited the sick fearlessly and ministered to them continually, serving them in Christ. And they died with them most joyfully, taking the affliction of others and drawing the sickness from their neighbours to themselves and willingly receiving their pains. And many who cared for the sick and gave strength to others died themselves, having transferred to themselves their death.
>
> (Eusebius 1890, 22, 7–10)

Abnormal diets may have had longer-lasting effects. As mentioned, the Fore people of New Guinea suffered a degenerative disease of the central nervous system though eating the brains of dead relatives.

Severe fasting may cause disablement. The Ethiopian Copts cause debility in young children when they fast for 110–150 days annually (Knutsson and Selinus 1970, 1956–9). Many pre-industrial communities have food taboos but since they are distributed over a variety of foods any ill effects are lessened. The impact on families is reduced if different members belong to separate clans.

The Moslem fast of Ramadan poses no health risk. There are a number of reasons for this. It is limited to daylight hours. Its long-term impact is reduced by following the lunar calendar in migrating through the solar year. It excludes particular groups: the old, the sick, travellers, pregnant women, nursing mothers, and those engaged in war. The worst problems are probably lack of fluids when it coincides with hot weather and a tendency to overindulge in large meals outside daylight hours. There are problems in high latitudes when Ramadan comes in the summer months and Moslems cannot even take required medicines. Long fasts can affect the environment in subsistence societies by reducing outputs when they coincide with times of peak demand for agricultural labour.

In many cultures, particularly non-monotheistic subsistence ones, the value of a person's life and its maintenance relates to his or her potential and actual contribution to the survival of the group. Many such societies killed a proportion of children, usually girls, who would have been a handicap to their annual survival. There is no point in keeping a defective or girl baby alive or prolonging the lives of the elderly if it reduces the survival chances of the younger members. This view is strengthened when the religion enjoins the continuation of an ancestral line and the ancestors are believed to keep an eye on their descendants' ability to keep the family going.

Ancient Carthaginian society was wealthy and well developed but sacrificed as many as 20 per cent of their first-born children to their god, Baal Hammon (Weyl 1968, 69–78). The Nayars of southern India were reported to have sacrificed their first-born to the goddess of smallpox (Sherring 1872, 81). These killings may well have had long-term effects on the environment as first-born children tend to become the more successful members of their societies. Eskimos and Bushmen have often abandoned their elderly as being just too expensive in terms of the others' efforts to keep alive (Carr-Saunders 1922, 223; Nansen 1893, 151). A distorted sex ratio among the Netsilik Eskimos would have posed a survival risk, compelling the hunting male who headed the family to provide food for a large number of consuming but non-hunting mouths. Therefore they killed 27 per cent of babies at birth (Rasmussen 1931). The Yanomamo tribe of Brazil kill the third child of either sex if the mother has two others who are still young. They also kill extra females on the grounds that a nursing mother would not be able to keep them alive and all would die if they tried to keep the newcomer alive (Chagnon 1977, 75–6). It is also probable that in the modern refugee situation where international aid has not been provided and in natural disasters such as the recent flooding in Maputo, the youngest children may have been abandoned for much the same reasons. In all these situations the religious ideas of those involved supported these actions on the ground that the survival of the group outweighed the interests of the individual.

Female infanticide also occurred as a result of the Hindu caste structure. Hindu India was and is an agricultural society with overpopulation problems. The religious emphasis on male heirs was the religious sanction for infanticide. Astrology also played a part. This often resulted in a surplus of women at the top of any sub-caste for whom appropriate husbands could not be found or at least not found without prohibitive expense. The result was that the society tolerated female infanticide (Dumont 1970, 111; Dickemann 1979, 163–95) and neglect of the health care of girls (Dubois 1972, 321–67). This has had some effect in controlling population growth which was seen as necessary to avoid overtaxing the environment. The 1852 census of Mysore showed that the Thakoor sub-caste had 10 695 male and 5865 female children, which outraged British administrators (Pakrasi 1970, *passim*).

Pastoral patrilineal societies such as the pre-Islamic Arabs with clan deities needed males to continue the agnatic ownership of herds and the need for males increased with the size of their herds (Smith 1901, 47). Status became a factor in Islamic times. Female infanticide reappeared when the number of men had been reduced by war and murder (Darlington 1969, 350), in spite of the Koran's condemnation of the practice. It states: 'the female child buried alive shall be asked for what sin she was put to death' (Koran 6: 141 and 152). In fact, the monotheistic religions have never favoured any killing of surplus mouths to ensure tribal survival. Both the Bible and the Koran are concerned with orphans. They are mentioned 23 times in the Koran and 41 times in the Old Testament but only once in the New Testament (Cruden 1769). Mohammed's sensitive view of life may have been conditioned by the death of his father soon after his own birth (Cragg 1971, 28).

9.3 Care of the short-term disabled

All pre-industrial societies have a substantial pragmatic understanding of ill health and the available means to deal with its particular forms. This understanding is often associated with the spirit world. Specialists and even ordinary people know large numbers of medicinal plants. It is probable that without such understandings some of these societies could not have survived. In East Africa the nomads knew that there was a connection between tsetse flies and the death of cattle and so they moved them at night, often with smoking torches, when the flies would not be so active. Some of the medicine which pre-industrial people discovered by themselves, such as quinine for malaria and rauwolfia for blood pressure, have become part of Western pharmacopoeias.

But although tribal peoples understand that one falls out of a tree because the branch breaks or that they suffer when bitten by a snake, they do not see such occurrences as accidental or having purely physical causes. There has to be a reason for what happened to that individual at that time and place and to no one else. They therefore differ from Western cultures in seeing the solution of such problems as being religious. Almost every individual wears amulets to prevent misfortune and encourage good fortune. This instinct is widespread in humans and exists in Oxford and Orlando as much as in Orissa or Onitsha.

When a disease or disaster affects numbers of people, it becomes a communal concern. Drought triggers the rituals for rain making. An epidemic calls on the combined expertise of a community. But it must start with individuals. Because they believe that illnesses and accidents have a supernatural cause, they call in religious specialists to neutralize the malevolence and restore the health of the individual.

Families in subsistence societies usually look after their sick at home, though they often invoke treatment by a specialist with religious motivation. Some of these have had high success rates but where there is no success, the sufferers continue searching for second and more favourable opinions. But individuals are not always able to afford such help so morbidity and mortality rates are higher. Only the small groups of people who have a direct interest in the patient attend the rituals to return them to good health. Nevertheless, the process of consulting a specialist credited with spiritual special healing abilities may work a cure whatever the details of the medicines prescribed. Such specialists would soon go out of business if they did not have a profound knowledge of their own culture and available local remedies (Lambo 1971, 8–9).

The care of the mentally disturbed requires systems of care with a spiritual basis. Diagnosis relates this basis to the disturbed state and involves the spiritual commitment of the attending professional. The schizophrenic has a better chance of recovery where his or her acts are seen as having a moral origin, than where, as in many Western treatments, they are seen as symptoms of a purely mental disorder (Field 1960, 53).

Most societies have religion-inspired institutions to care for the sick and needy. These include special staff able to act independently of some at least of their culturally accepted familial and communal obligations.

Medieval Hinduism had religious institutions carrying out medical work for the indigent without having hospitals in the modern sense, but little is known of them apart from inscriptions (Reddy 1941, 385–400).

Islam has donations to charity as one of its five fundamental requirements and substantial funds known as 'waqfs' have always been available for charitable purposes. By the end of the seventh century hospitals rather than hospices became a feature of all the main cities of the Islamic world (Hamarneh 1962, 336–84). The larger ones had medical schools attached to them and the care of the sick was systematic in wards with physicians keeping records (Elgood 1951, *passim*; Al-Kindi 1966). Since Islam did not impose religious restrictions on care, this was able to proceed unimpeded, particularly as the doctors were not clerics.

Christianity has historically related illness to sin and devil possession (Ackerknecht 1971), with the priest working beside the doctor who was seen as the servant of God (Amundsen and Ferngreen 1986, 113–19). In modern times secular governments have increasingly taken over the functions of religious hospitals and other health institutions. But the extent of this development varies between countries. Particularly in the United States there are large numbers of hospitals run by Christian and Jewish charitable trusts in which the religious faith sets the tone in which their work is carried out.

In many developing countries, Christian missionary organizations have been largely responsible for building and maintaining welfare institutions. They have provided a major part of hospital and dispensary services sometimes without official subsidies. These efforts have had substantial effects on the survival of nearby populations. Roman Catholic organizations tended to predominate because of their numbers and the dedicated personnel in the missionary orders. The White Father missionary order in the Congo had 16 hospitals treating 200 000 patients annually in 1911 and the Holy Ghost Fathers had 176 hospitals and dispensaries in tropical Africa as a whole in 1931. Roman Catholic missions alone were running 120 leprosaria treating 17 000 patients by 1940. The Franciscan Missionary Sisters of Africa in Nsambya, Uganda, began building their hospital in 1906 before they had built a convent for themselves (O'Hara 1980, 27–8). The Church of Scotland missions in Kenya required each station to have a resident doctor (Philp 1936, 83), while some groups such as the Universities Mission to Central Africa paid less attention to health including their own until well into the 1920s (Ranger 1981, 261–77).

There was a change in emphasis in medical missionary work by Western church groups during the twentieth century, deriving at least in part from the wide acceptance of the Darwinian perception of the origin and nature of man. In the nineteenth century religious conversion had generally been seen as the necessary foundation for practical advances in medicine and the economy. Missionary effort had aimed primarily at conversion with medical help as a secondary adjunct. In the twentieth century, partly through the influence of Albert Schweitzer (1963, *passim*), the emphasis has shifted to medical and economic aid as the root rather than the fruit of religious advance. This has enormously increased the focus on medical and agricultural aid through agencies such as Christian Aid, Tearfund, CAFOD (Catholic Aid) and others. These have been responsible for the survival of many thousands made destitute by war, famine and natural disasters.

Missionary medical work and emergency aid have had two effects on the environment. First, they have improved the health work and raised the survival rate of particular populations such as the young, women, the elderly and the disabled, who would almost certainly have died. Secondly, they created colonies with specific populations around their institutions.

9.4 Care of the illegitimate, orphans and the long-term disabled

The care of the illegitimate, orphans and long-term disabled affects the environment by increasing their lifespan and mobility, and thus requiring the provision of accommodation and supplies. The monotheistic faiths have always been concerned with orphans and abandoned children. From the earliest days Christians, notably religious orders, have run institutions for foundlings, but their success rate was often very poor. The death rate of foundlings in eighteenth-century Montreal was 92 per cent (Fortier 1963, 36, 351–9), possibly partly because of the difficulties of feeding them before the invention of vulcanised rubber for the feeding bottle in the next century.

The slur of illegitimacy also handicapped their survival chances. The death rates of illegitimate children have nearly always been proportionately higher than others, regardless of social factors (Haney 1972, 303–10; Scurletis et al. 1969, 214–21).

Judaism and Christianity have always required the segregation of lepers, the latter requiring them to be maintained at church expense. In the Roman Catholic Church they were under the care of their bishops (Mercier 1915, 8, 27–8). Lazar houses were established throughout Europe in the Middle Ages to isolate sufferers. In the fourteenth century there may have been as many as 7000 in France alone and some of the earliest hospitals in England were established for lepers. Most missions run similar institutions in Third World countries today so that the overall numbers of lepers treated is considerable.

Overall it is the Christian churches which have consistently attempted to care for the abandoned in the societies in which they operate even after some of these functions were taken over by governments. By 1931 the White and Holy Ghost Fathers in the Congo had 152 orphanages and in 1981 the Salvation Army in Kenya alone had four institutions for the handicapped and six for the blind, including a secondary school which achieved high examination results.

The funds which support the religious care of millions of people with long-term disabilities are now mainly from governments but are significantly aided by church-funded charities. Their care involves the erection and maintenance of communal buildings, arrangement for the transport of supplies, and the accommodation of professional and support staff. In private homes, it may involve the modification of rooms, staircases and bathrooms, and the provision of facilities for medical visitors.

In societies with a low expectation of life, those who survive beyond their thirties or forties often become useful elders and statesmen, a point which Judaism stresses (Joseph 1917).

10
Religious Influences on the Life Cycle and the Maintenance of Environmental Balance

10.1 The chances of conception

Monasticism and the celibacy of priests, monks and nuns clearly reduce the rate of conception in a community and may even determine whether a particular population expands or contracts. Buddhist monasticism has similar effects. In Outer Mongolia the number of celibate lamas has been associated with the slow growth of the population (Dondog 1972, 52–4). Some religious beliefs have led to practices which have terminated the possibility of conception. The upper-class male members of the Skopsi in nineteenth-century Russia castrated themselves. This sect, which started in 1770, expected the millennium when their numbers reached 144 000 as mentioned in the book of Revelation (7: 4; Conybeare 1921). Similarly, the separation of male and female Shakers precluded births. Those who wanted to conceive would have left their communities. However, since female members were often those seeking refuge from male mistreatment, this may have been infrequent.

In some cases a combination of socio-religious factors may have kept the birth rate low where mechanical birth control was not an option. Many believed their ancestors wanted their families to continue, so they restricted births to provide the children who were born with a better chance of survival. Most pre-industrial societies have prolonged breastfeeding since there are no special infant foods, and they maintain a social embargo on sexual intercourse during this period to prevent a further pregnancy which would jeopardize the existing infant. The Koran specifically encourages breastfeeding (Sura 2: 223). This practice delays the resumption of menstruation so whether or not intercourse takes place, the chances of conception are lower than for women who do not breastfeed their babies. The post-partum taboo on sexual intercourse among the Mekranoti Indians of Brazil was 36 months prior to 1955 and fell to 25 months after 1965 (Werner 1983, 227–45).

Hinduism has a dichotomy of approaches towards sexual activity. On the one hand, wandering sadhus generally take vows of poverty and celibacy, and depend on householders for their support. Towards the end of life ordinary Hindus are required to detach themselves from worldly concerns, although this applies as much to diet as to sexual abstinence. The devout Hindu, in conformity to phases of the moon and especially to religious days, might well refrain from sexual activity for 120 days annually (Chandrasekaran 1952, 73–9). Such abstinence, however, is less common among the young. On the other hand, much Hindu art is highly sensual, as exemplified by the sexual acrobatics depicted in the sculptures on the outside walls of the temples at Khurajaho.

One cause of low birth rates among the higher-caste Hindus is the age disparity between husbands and wives who historically were married off before or very near to their menarche. Thus many were widowed and unable to remarry while still virgins or even in their teens. So Hindu India has always had large numbers of higher-caste widows for whom remarriage is impossible. It was less difficult for those lower down the religious scale because they were needed as farm labourers (Dandekar 1959). Many pre-industrial cultures allow for the widow to be inherited by a brother-in-law as the original marriage made her the legal property of her husband's family. Judaism also permits levirate marriage including the right of the brother in law to allow the widow to marry out. Christianity has no restriction on the remarriage of widows and Islam permits it once the widow has menstruated four times.

Buddhism regards celibacy as an ideal. It discourages any form of attachment to worldly affairs from indulgent eating to sexual intercourse, and it luridly describes the latter as resembling coitus with a snake's mouth. All men, including those of the Thai royal family, spend some time as monks. Well over 5 per cent of the fecund male population are celibate at any one time. Widowed Burmese, for instance, show a disinclination to remarry. However it is possible to buy very explicit fertility amulets within the precincts of Bangkok temples.

Islam has no celibate ideal even for acknowledged mystics. It positively encourages sexual intercourse in marriage, specifically at night following the daily Ramadan fast. There are no periods in which

Moslems are expected to refrain from sexual activity for religious reasons except during the once in a lifetime pilgrimage to Mecca. Islamic polygyny probably originated in the need to care for the widows of battle casualties in the early days of the small Moslem community. But any resulting increase in the birth rate may be somewhat lessened by the accompanying restriction to a maximum of four wives. It may be still further lessened by the requirement that the husband treat all four wives equally in all ways including sexually. Some have contended that the difficulty of this meant that Mohammed was in effect advocating monogamy. But it does allow the husband access to another wife when one is breastfeeding. Nevertheless, women in monogamous unions do tend to have more children than those in polygynous households. A comparison of age-specific fertility rates for monogamous and polygynous women showed that at every age the rates for monogamous women were substantially higher (Ukaegbu 1977, 397–404).

But on balance, polygyny does increase the overall fertility of Islamic communities and has led to rapid population increases in Moslem countries. Although the majority of Moslem marriages are monogamous, having additional wives is prestigious although expensive. It means that the husband with a second wife is likely to be older than a man marrying for the first time and his potency lower. This must also be a contributory cause to the instability of Moslem marriages and explain the frustration felt by some wives which leads them to seek spiritual support. This can open the way to spirit possession. It is widespread in Islamic societies, notably in the Sudan (Kenyon 1995, 107–20) and in the zar cults of the East African coast (Giles 1987, 234–58). It can perhaps be seen as a way in which a marginalized wife gets her own way. This is because once she is possessed, her behaviour is out of her own control and it is the possessing spirit that has to be appeased enough to leave her. It can be expensive for the husband in having to pay for gifts and ceremonies (Behrend and Luig 1999). Islam's rules of inheritance which give women specific proportional rights to the property of their deceased parents and husbands adds a commercial flavour to marriage since it makes it likely that a husband can inherit his wife's family's property on her death. Thus in Islam a high proportion of men marry their first cousins to whom they are related through both their fathers and mothers. The high fertility of Moslem communities is a matter of environmental concern when it exceeds that of comparable non-Moslem communities with which they share the land. This is the case in the Lebanon where the Moslem birth rate is higher than the Christian. It contributed to the recent civil war as constitutional equality was seen as unbalanced when census results were published. It is also a concern in Israel where the birth rates in 1985–89 of Jewish women was 2.6 and that of Moslem women 3.49 (*Encyclopedia Judaica Yearbook* 1990–91) leading to increased Jewish immigration and greater negotiating intransigence.

Polyandry is much rarer than polygyny but is characteristic of some Himalayan societies. Several men, usually brothers, may have one wife. This would be expected radically to reduce birth rates. However, this does not seem to happen in Nepal where Tibetan Buddhists do not have a substantially lower fertility than Hindus. However, the lower rate for the latter may be partly due to restrictions on widow remarriage. Also Tibetan polyandry is usually two brothers saving on costs by having a single wife, but as one brother usually has priority, the overall effects are the same as monogamy (Ross 1984, 163–81).

All Christian denominations have sought to confine sexual activity to married couples while ultimately recognizing that this ideal would not always be realized. Christianity does not allow divorce or annulment on the grounds of infertility. This has sometimes induced farmers, in order to have children to help on the farm, to postpone marriage until they know that their prospective wives are pregnant. There is also evidence that they timed births to avoid wives being unable to help with key agricultural periods (Arbellot 1970, 225). The highly fertile Hutterites in this century seem to have timed births to coincide with lighter work periods (Surbey 1986, 337–45).

From early Christianity onwards there has also been a celibate ideal to which many have become institutionally as well as personally committed, thus freeing them to help others and to concentrate on their devotions. In the Middle Ages, when male mortality was high from war and monogamy was a social requirement, very large numbers of women became nuns. Male commitment to celibacy may also have fluctuated. The religious life was a protection against secular hardships, but its attractions diminished in England after the Black Death and the rise in the value of working men. However, for long periods in the Middle Ages in Europe some 5 per cent of the fecund population may have been committed to this celibate religious life. In Siena, Italy, in 1670, 11 per cent of the total population were either monks or nuns (Beloch 1937, 76–7) and in Tudor England the secular priesthood amounted to 4 per cent of the total male population (Haigh 1993, 5–6).

Further constraints on sexual activity came from the Church calendar. Medieval Christians were expected to refrain from marital intercourse on Saturdays and Sundays, during Lent, on important religious feast days, and on the night before taking communion. This final injunction remains to this day (Puzo 1972). This gave a total of some 160 days of abstinence each year (Bieler 1963).

The Roman Catholic Church at one time feared population reduction because of the tabooed sexual relations of monogamous men with their breastfeeding wives, particularly when withdrawal before ejaculation was disallowed by church rulings, whereas it was advised for Jewish men throughout the nursing period (Noonan 1966, 69–76). The Church supported the practice of wet-nursing by another woman, thus providing 'for the frailty of her husband by paying the conjugal due, for fear that he might lapse into some sin against conjugal purity' (Fromageau 1733, 1198). Data from seventeenth-century France showed that the birth interval for those women who had put their children out to nurse was shorter than for those who did not (Lachiver and Dupaquier 1969, 1399). On the other hand, this practice may have as much as doubled the infant mortality for the same period (Bardet 1973, 28–9).

Christianization has had relatively little effect in reducing polygyny in West Africa despite considerable missionary activity for over a century. There is little difference between Moslems and Christians. Half the men and two-thirds of the women had been in polygynous marriages at some time in their lives (Orubuloye et al. 1994,1121).

The main effects of overpopulation resulting from unconstrained conception will be more intensive land use, urban expansion, the colonization of new areas and emigration.

10.2 The chances of survival into adulthood

Infant mortality has always been a serious problem and has a strong religious dimension. In general, children's lives are best preserved by the prosperity of the parents and the quality of life in the home. This includes the usual basics: good diet, exercise, rest, hygiene, avoidance of harmful substances, and domestic and communal harmony. These ideals have seldom been realized although the scientific and social advances of the last century have given a vast increase in infant survival rates. This has been more marked in Christian societies than elsewhere although vitiated by the frequency and seriousness of wars and social upheavals. Both the highest and lowest figures were from areas that would call themselves Christian: 89 per thousand live births in Africa as against 5 in western Europe (*Encyclopedia Britannica* 1998, article: Demography).

Illegitimacy affects the survival rate of children. They are more likely to be aborted and to receive inferior care. Their chances are best in developed societies, though this difference is decreasing. In Italy in 1950 the differences were very marked, with the legitimate death rate being 62.3 and the illegitimate 107.8 per thousand live births but this had evened out to 36 for both by 1965 (Nodari and Pirovane 1970, 1254–6, 1449–51). Religious action has, however, always mitigated the greater risks suffered by illegitimate children. In pre-Reformation England, for instance, where the priest's livelihood depended on the goodwill of his parishioners, the concept of illegitimacy may have been elastic and humane (Marshall 1994, 19–20).

Islam has been particularly concerned to remove the stigma of illegitimacy from innocent children. Its laws have therefore ignored the length of pregnancy and declared children legitimate when born to men who are dead or who divorced the mother years before, if she has not remarried. The Shafi and Maliki schools of law teach that a man can acknowledge a child as legitimate if it was conceived before a marriage or up to four years after the marriage has been terminated. The concept of illegitimacy is thus very largely absent from Islam, but this of course is within a culture which segregates women and greatly restricts their social contacts.

Legitimacy is not always straightforward. Some cultures in New Guinea believe that conception is related to the activities of ancestors even where the woman's husband may have been away as a migrant labourer for longer than any possible pregnancy. The Nuer have 'ghost' marriages for men who have died without male heirs. They permit a girl to be married to his ghost and then conceive through a lover. The child on birth becomes the dead man's legitimate heir (Evans-Pritchard 1956). This has also happened among the Kamba in Mombasa where a successful businesswoman, not wanting to have her affairs controlled by a husband, married another woman who conceived via her lover and the child was considered to be the first woman's legitimate heir as she had followed the customary procedures for initiating a legitimate marriage. During the First World War the French government provided for men who were killed but who had had an accepted relationship with a woman who became pregnant. They allowed her in law to be married and pensionable and the child legitimate. In all these relationships religions and cultures have tried to regularize family structures so that children will have a better chance of survival. Even where the birth is irregular they award them a socially acceptable status.

There is sometimes an increased death rate among children born to legitimate unions. Cases exist where children have been killed for religious reasons. Examples are the killing of twins and those born by breech presentation (Schapera 1966, 223–5) or whose upper teeth emerge first (Cory 1951). The whole question of twins worries some pre-industrial subsistence societies where such multiple fertility gives the parents an advantage, while making the survival of such children problematic. The baptismal registers of Roman Catholic churches in Ngara, Tanzania, show a high proportion of twins baptised in expectation of death, and thus allowing the parents to avoid non-Christian obligations to hold special and expensive ceremonies.

A society can lose large numbers of boys, but it is the survival of girls which maintains population levels and leads the recovery from disastrous wars and massacres. Belief in horoscopes can affect this. Japanese girls born in the Year of the Horse were thought to make poor wives. The result of this was a marked leap not only in abortions but in the neonatal mortality of girls. In 1965 the rate was 5.17 per thousand live births but in 1966, the inauspicious year, this rose to 7.78 with no comparable rise in the deaths of boys (Kaku 1975a, b).

Most religions and societies have male-centred ideologies and prospective parents often give better treatment to boys after they are born. Hinduism, for instance, underpins this inequality since it has nothing concrete to say about the ways to feed girls and they receive medical treatment after boys. The reason is that sons bring in money to their families when they marry while girls cause poverty through requiring a dowry. On the other hand girls are far more useful in conditions of poverty than their male counterparts.

The Hindu caste system gives children born to lower or Outcastes less access to the means of survival and they have much higher infant mortality rates. In the 1940s lower-caste Hindus in Bombay had an infant mortality rate of 272 per thousand live births, upper castes 208, Muslims and Indian Christians 190 and Parsis 88 (Chandrasekhar 1959,105). While the rate has been reduced, the differential continues partly because population growth has handicapped improvements in post-natal services and standards of living. Hindu confinement is ritually unclean and thus midwives have been of low caste, both illiterate and ignorant. This lack of hygiene and ritual impurity increase infant mortality. The segregation of women in both Hinduism and Islam does not encourage midwifery as a profession or confinement in institutions apart from the family.

In contemporary China the national census has disclosed a marked imbalance between the sexes as a result of their 'one child' policy which has brought a large-scale infanticide of girls. Without the abhorrence of infanticide induced by religion, handicapped children are disposed of without moral scruples in many subsistence economies. Thus a Sukuma blind girl would be kept alive because she could be useful around the house and her fecundity is still socially marketable but at a depressed level, while a blind boy might be neglected and die. In Europe, economic stress sometimes gave a bias towards the care of daughters (Voland 1988, 392–400) and among the Sukuma it was consistent with custom to marry daughters to men with food in times of famine (Cory 1953, 50).

Christian societies accept the survival of the handicapped child as a gift from God and a special responsibility (Zuk et al. 1961, 525–40). They use personal and communal resources according to the degree of the handicap although this inevitably reduces what is available to others. Missionaries and missionary orders have often led in the care of abandoned children and orphans but the numbers have always been relatively small. They have also done their best to provide for the victims of war and natural disasters. But young children often die before help can come since the places of succour are sparse and often inaccessible. The stark alternative is often walk or die. There is often very little that religions can do other than stimulate the well-to-do in other societies to provide funds and aid personnel.

The mother's legal situation in giving birth also affects the child's survival. In difficult births, Judaism gives preference to the mother (Levine 1968, 638–9), and Islam and the Eastern religions conform to this view. The Roman Catholic Church, on the other hand, teaches that in cases where only one life can be saved, it should be that of the child (Capellmann 1882).

There are other religious rulings which can reduce the chances of children surviving and thus reducing pressure on particular environments. Some Christian sects have objections to immunization. This is most strikingly visible in the USA which has a large number of small religious groups with beliefs and practices divergent from the mainstream of American society. All states except Virginia and Mississippi allow exemption from immunization of schoolchildren based on religious beliefs (Novotny et al. 1988, 49–54). This may affect survival rates where such groups are concentrated (Frank et al. 1985). In 1985 two groups of Christian Scientists experienced outbreaks of measles, 90 per cent of the cases occurring among those considered religiously exempt. The fatality rate was 2.2 per cent as against a 15-year average rate for the whole of the USA of less than 0.1 per cent.

10.3 Marital arrangements

Marital arrangements significantly affect population growth. Religions generally agree that an accepted relationship between a man and a woman and the resulting children gives a child a better chance of survival than irregular unions can. As part of this process parents exercise constraints on the possible spouses for their children even though the children themselves may well steer their parents' selection in particular directions. These choices tend to be both religiously and socially endogamous. Calculations rather than emotions sway their judgements. The famous Indian writer Nirad Chaudhuri married late in life through the choice of his mother and knew little about his wife until the first night when he found that they had a shared love of music which he described as a good start. Different faiths, however, differ in the emphasis they place on monogamy and this affects population growth. Buddhism, since it has no god and dissuades people from any form of addicting relationships, is not involved in marriage ceremonies except as an occasion for the feeding of monks. Islam sees marriage as a civil contract and not as a sacrament. Although the Prophet Mohammed enjoined good treatment for rejected women, divorce is relatively easy and frequent. Among the Shi'ites there is even a form of contractual marriage for a fixed period which can be as little as a week or a university term. Hindu wives are enjoined to see their husbands as quasi-divine and the marriage indissoluble even after death (Agrawal and Agrawal 1972, 73–80) and Christianity similarly sees marriage as a sacrament and as indissoluble except for adultery.

Age at marriage affects fertility for both men and women. Hinduism endorses early marriages with the Laws of Manu stating that girls aged eight years were suitable for marriage with a Brahman. This does not, however, necessarily affect the birth rate partly because women do not reach optimum fertility until some years after the menarche. Judaism and Islam have also favoured early marriage for girls but this is in the context of strong family support. Such support is often lacking in Western society and can cause strain when young couples set up households on their own. The resulting stresses, often accompanied by family tensions, usually restrict family size. In the Hindu joint family and the less patriarchal Chinese and African family structures, such tensions usually originate between the mother-in-law and the incoming daughter-in-law. This has led to suicides in China and murder in India where families may have been seeking higher dowry from a replacement wife for their son. The permitted age differential between husband and wife which is particularly marked in polygyny leads to older men with much younger wives and a corresponding lower potency. Recent mass Roman Catholic weddings in Uganda have had these marked age gaps because the younger women are considered to be free from AIDS and the older men have the means to tempt them into marriage. Also, the older and almost certainly richer husband is in a better position to ensure the survival of any children which are born.

Some Jews believe that men have a duty to continue sexual intercourse with their wives after they have the desired two children, one of each sex (Dorff 1986, 25). The Mishnah (Ketubbot, Torah, 5: 6) spells out the correct frequency of marital relations which should be once a day for a man of independent means and once a month for camel drivers. These injunctions have probably helped to sustain the Jewish population.

Divorce reduces population by reducing the birth rate. Its trauma can affect concern for the environment. With small exclusive groups, divorce is likely to be limited. The Amish with their pattern of living close to each other, combined with self-help, tend to favour cousin marriage, 47 per cent in a recent study (Hurd 1985, 82–9). In southern India, Hindus have the same pattern of inbreeding with the addition of uncle–niece marriages (Rao 1983, 413–22).

The Hutterites have communal feeding and only a single room for each married couple. However, they do not have the same problem of inbreeding. When numbers necessitate, they form new colonies which can be some way away and beyond horse-drawn calling distance. This makes the new and old communities genetically distinct. They realize, however, that taking up too much land in any locality will give them social and political problems. They have therefore gradually expanded what they consider to be the optimum distance between their colonies from 30 to 100 miles (48–160 km). This causes no problems for their way of life since each colony is self-supporting. They only divide when the numbers in a projected new colony are sufficient to secure its independence. By this system of distancing, they avoid having to compromise their objections to mechanical birth control.

A further restriction on birth rates is delayed marriage. Girls in Hutterite farming communities prefer long courtships. They want to enjoy the advantages of their own community for as long as possible before moving to their husbands' colony where they will be strangers. Also, colonies which have a poor reputation from drought, mismanagement and poverty have difficulty in getting wives for their young men (Hostetler and Huntington 1967, 102). However, the Hutterites have steadily expanded their numbers and have been buying up new farms. But the younger men are beginning to recognize that there has to be a limit to their process of unrestricted growth by which their wives have in the past averaged seven children each. However, these groups on the whole have grown until they reached the capacity of their lands.

In the nineteenth century Mormons were allowed to practise polygyny to maintain their numbers in face of a greater death rate among men. This has had a very significant effect on raising the population. A study showed that men with one wife averaged 7 children, men with two wives averaged 16 children, and those with three wives, 20 children. Polygynous Mormon men as a group averaged 2.4 wives and 15 children while polygynous Mormon church leaders in particular averaged 5 wives and 25 children.

Female circumcision, as practised in parts of Africa and the Near East, is a cruel sewing up of the female external genitalia and excision of the clitoris. It eliminates those of delicate health and reduces the woman's sexual enjoyment. Its object is to reduce the woman's temptation to adultery and, it is said, to increase the sexual enjoyment of the male. Its occurrence among Australian Aboriginals appears to correlate with areas of poor rainfall. It occurs widely in Somalia and the Sudan. It makes the birth process more difficult and so may well contribute to infant mortality.

10.4 The disposal of the dead

The disposal of the dead is a charge on any society and there are many institutionalized procedures for achieving it. They are almost always based on religious considerations and regard for the feelings of survivors.

Very few societies dispose of bodies in a naked state. They usually process them in ways that at least show that they care for the deceased and have some regard for their status and religious beliefs. But methods of corpse disposal have tended to be utilitarian in the extreme. English law in the Middle Ages required people to be buried in woollen shrouds in order to support the nation's main export.

Religion has always prompted mankind to attribute a special value to the human body. This has prevented corpses being viewed simply as a resource. Public opinion was appalled when Nazi Germany used parts of the bodies of concentration camp victims and extracted their gold teeth. On the other hand, the cadavers of animals, even if they are not slaughtered for human food, are utilized as animal food, fertilizer, leather, glue and fuel.

The most environmentally friendly method of disposal is burial where the body's decay contributes to the regeneration of the soil. Sometimes this is recognized as in the phrase traditionally uttered after a death in Crete, 'this very earth which nourished you will eat you as well' (Lioudaki 1939, 46 quoted by Danforth 1982 plate 13). Some societies do not have tools strong enough to dig graves in hard soils, particularly in the dry season in tropical Africa. In such circumstances they cannot bury the dead deeply enough to prevent infection. The traditional 'honey-pot' posture for burying the dead among the Sukuma of Tanzania is occasioned by this difficulty.

Parsis have a method for disposal of the dead which assists the survival of other animals. Pall bearers carry the dead body to a 'Tower of Silence'. In Bombay these are erected in a beautiful garden on the highest point of Malabar Hill, amid trees swarming with vultures. The towers are constructed of stone, about 8 m high with a small door at the side for the entrance of the body. It is then laid on the tower on an iron grating (Figure 10.1). When vultures have devoured the flesh the bones fall through it. On the third day after the death the relatives perform a

Figure 10.1 Parsi Towers of Silence in which the dead are abandoned

ceremony of remembrance and dedication in a 'Fire Temple' where there is a perpetual flame.

The Masai, a cattle-keeping tribe in Kenya and Tanzania, also leave their dead exposed to be eaten by scavenging birds and animals. This is one reason that they will not kill hyenas who may have been involved in disposing of the bodies of deceased relatives. This custom may also be related to their cultural or religious dislike of cutting into the earth, and to the very practical fact that they do not have the tools for digging since they are not an agricultural people.

The use of the body for medical research and of its organs to keep other people alive is a recent contribution to a benign environment for humans, but there are religious considerations both enabling and discouraging such uses. Judaism holds that respect for the dead as God's creation prevents bodily mutilation for dissection or post-mortem examination unless it can be of immediate value in saving the lives of the currently ill. Christianity generally permits both without serious restrictions. Islam forbids the disfigurement of the dead but has not opposed their limited use for medical research. This is not a problem for Buddhism, Hinduism or other eastern religions since they tend to see life as transitory and repetitive (Geller 1984, 494–6) and the body as temporary housing for the transient spirit and therefore disposable.

The largest effect on the environment from the disposal of the dead is the use of land for graveyards and cemeteries. Traditionally such land is regarded as consecrated. The quality and adornment of an individual grave depend on the wealth and status of the one buried. England has numerous long barrows from the Stone Age and thousands of round barrows from the Bronze Age but they represent only a minute fraction of those who died. Since Roman times graves have been marked by stone or metal memorials rather than mounds and this custom has been followed by Jews, Christians and Moslems. Tombs are often an important feature of the environment. Ancient Egypt is perhaps the outstanding example. They placed the mummified bodies of pharaohs in tombs carved into the rock and in about 80 pyramids along the west bank of the Nile with the largest towering over the desert near Cairo. These used up vast amounts of materials and both skilled and unskilled labour, but the graves are in useless desert land, as were the pre-Columbian burials in coastal Peru.

Today, the Egyptian pyramids and tombs are tourist attractions supporting a very large number of workers whereas the support for their original construction must have come from the royal treasury. The diversion of labour from agricultural work and the extension of cultivable land along the Nile valley must have been a severe economic burden. It has been estimated that the Great Pyramid of Cheops at Gaza required 100 000 workers over 30 years and exhausted Egypt more than a major war would have done.

Where space is at a premium it is often necessary to reuse the same grave for subsequent burials. The cemeteries of English parish churches which have been cut through by road widening or urban development, often disclose layers of skeletons possibly buried 100 years apart. This superimposing of burials may have been common in the past and inside churches as well as in churchyards. A Dutch painting by Emmanuel de Witte of a Protestant funeral (Boymans-van Beuningen Museum, Rotterdam) shows bones and skulls from earlier burials beside the open grave in the choir of this church.

Much the same occurred in Cretan and mainland Greek villages where the bones were dug up after the five years of mourning, washed in wine and placed in ossuaries. As diggable rockless ground was scarce and needed for crop growing they had to reuse the grave site. Even the marble monument itself could be used for the grave of the next villager to die (Danforth 1982, 14–15). Cemeteries in even small isolated villages are now filled with permanent grave memorials which prevent the sites from being used again or expanded.

The alienation of arable land for use in burials was a problem in ancient China where they also constructed burial mounds. They considered that cremation showed great disrespect to the dead. A petition presented to the Board of Revenue in about AD 1280 (Moule 1957, 45) refers to the high cost of burial bringing poor families to ruin and the need for magistrates to make extensive burial grounds available near to the town on land which otherwise would be unusable. At that time land was scarce and fuel cheap but by the time of the Ming Dynasty (AD 1368–1644) the continuous felling of forests had made fuel scarce and land more plentiful so that burial became commoner again. In many areas, the use of timber for the casing of the body or its cremation has resulted in deforestation and erosion of denuded slopes.

The Peoples' Republic of China has drawn attention to the enormous consumption of materials in funerals. Grave mounds in the middle of fields are common near to Peking and Xian. When most farmers cultivated by hand, they could plant crops without difficulty up to the sides, but they are much more of an obstruction to tractors. A Communist official in Shantung Province stated that 3 per cent of arable land was occupied by graves and tombs, the burial clothing of the deceased annually used up 24 million feet of cotton cloth and the lumber used in the annual construction of coffins would build 200 000 houses. This much reduces potential food, housing and clothing for a still expanding population (Macinnis 1973, 191).

The Chinese government in Taiwan has also found the customary disposal of the dead and the accompanying ceremonies to be both wasteful and polluting. Funerals are accompanied by the burning of papier-mâché models of everything that the deceased might need in the afterlife. The resulting smoke which both contributes to air pollution and burns up a precious natural resource causes concern to the Taipei City Council.

But the Chinese still align and place burials in accordance with feng-shui. In the New Territories of Hong Kong this is elaborated into three stages: the coffin is buried for about seven years, then the bones are exhumed and left in a pot on the hillside, and finally for some the pot is permanently entombed (Nelson 1974).

As the standard of living rises in a society, more materials are used not only for funerary monuments but also in providing grave goods for the deceased's hypothetical comfort. In modern Christian or post-Christian societies all the dead tend to be buried in their best clothes or in clothes specially bought for the occasion. In Greece a woman dying before her marriage would be buried in her wedding dress (Danforth 1982,13). This expenditure on the dead has recently received considerable literary treatment (Waugh 1956; Mitford 1963), as well as being satirized in film.

Banaras is a good example of how the inadequate disposal of the dead can contribute to the contamination of the environment. The water is polluted by some bodies that are half-burnt, some which have been immersed without cremation, and the ashes of those who have been cremated elsewhere. Daily 40 000 visitors use the Ganges for ritual bathing despite the fact that human remains and untreated sewage contaminate the water to a level 24 times above that recommended for bathing (Alley 1994, 127–45).

10.5 The influences of the dead on the living

Although biblical Christianity rejects the idea of consciousness after death, people in all cultures speculate from time to time on what their dead parents would think if they could see them now. For most of the Western world this wears off when the period of psychological rather than social mourning is over.

Cemeteries protect land from commercial use. They are often so ill-kept as to constitute nature reserves. In the cities they can function as supplementary lungs, and as a relief from the monotony of housing. Several London cemeteries such as Kensal Rise are official nature reserves, while that in northern Highgate could be described as a forest reserve since it has been almost abandoned by the living.

The memorials for the dead are in many cases specially designed and expensive to erect. Sometimes the amount of care given to graves suggests that neglect sets in soon after mourning is over. The main Roman Catholic cemetery in Manila shows gross neglect, whereas the Chinese one next door is immaculate with each grave carefully tended and seemingly the scene of regular ceremonies of remembrance. Croatians carpet all their cemeteries with flowers on All Souls' Day with a profusion which suggests the involvement of specialized growers and nurseries.

The graves of some important people remain the centre of both secular and religious devotion for centuries after their deaths. They can become pilgrimage sites and devotees make repeated visits to them. There is a permanent queue to file past the tombs of Mao Tse-tung, Ho Chi Minh and Elvis Presley, perhaps partly to reassure the viewers that they are in fact dead. The areas surrounding such graves become sterilized against any other activity, often as a result of the provisions of the deceased's will.

Death often brings widowhood. The religious rules which surround it can affect population growth. Some religions enjoin mourners to keep themselves segregated after death. Hindu widows, who are often much younger than their husbands, have great difficulty in remarrying in families which do not need the labour of an extra pair of hands. Buddhist reluctance to remarry comes from the belief that it is inauspicious. As a result, widows and widowers often marry each other (Spiro 1977, 148). Islam has no objection to the remarriage of widows and polygyny prevents a surfeit of widows for whose maintenance no one is responsible.

An excess of impoverished widows occurred in Christian Flanders when many of their husbands were killed in private wars. Their relatives cheated them of their agreed dowry rights and estates and they had grown burdensome to their children. They flooded the church in AD 1070 so that it had more nuns than it could reasonably utilize and maintain (Kindere 1904).

'Ancestors' is a wide term and includes any antecedent. Some tribes believe they are conscious in death and can send distress or misfortune to the living whose behaviour displeases them. In patrilineal Africa diviners usually identify the cause of this misfortune as some breach of custom within the family or clan. It can affect their family's lands and livelihood

The continuing involvement of mainland Chinese with ancestor worship is not confined to backward villages with few Communist Party members. A majority of subsistence farmers feel obliged to make regular offerings to deceased kin, hoping for some beneficial influences from ancestral spirits. There has recently been a shift from collective to domestic rituals with memorial tablets. This has not much reduced employment because even in pre-Communist time Chinese religion did not depend on full-time temple priests. Since the government has confiscated family- and temple-owned lands, these ceremonies around the dead provide farmers with extra hope and a sense of belonging despite political changes. They are commonest in villages getting money from overseas Chinese who expect their ancestors to be appropriately remembered, and until recently in collective farms either because of their larger size or the predominance of a single lineage (Parish and Whyte 1978, 254–96). Much the same approach exists in Japan where ancestors are also of religious concern to the whole family (Takeda 1976, *passim*).

Death sometimes inhibits marriage for quite long periods. With the Lepchas of Sikkim this starts not only after the death of a spouse, but also after that of a parent or sibling (Gorer 1938, 361). Bereavement can also affect survivors by depressing their efforts for a period. In developing societies the loss of a mother is often more distressing than that of a wife because the latter has a less pivotal role in the home and is more easily replaced. In the Western developed and individualized world, by contrast, the loss of a partner may be more disturbing, giving the surviving partner a markedly higher short-term mortality and morbidity from a wide range of diseases (Maddison and Viola 1968, 297–306). This is an added environmental burden on any developed society (Rees and Lutkins 1967, 13–16).

10.6 The prohibition of contraception

There is always a doubt of the reliability of statistics on contraception. It is a private affair. Surveys must rely on individuals responding accurately to probing questions about their private lives. Roman Catholics questioned are likely to answer at least in part in accordance with what they feel should be their expressed views. Only one partner may have used contraception so that in terms of church discipline, the other may have known that it was being used without directly breaking the moral ruling.

In general, the use of contraception is thought to reduce population growth and hence environmental pressures. The connection between religion and population pressure is usually related specifically to the Roman Catholic canon law prohibiting artificial contraception. This doctrine has been in existence for many years and has clearly had important demographic effects. It contributes to the overpopulation of poor countries, notably in Latin America, where the poor have rarely used contraception. Whether they knew of the Church's ruling or not, they would almost certainly have followed it. Children are useful and fertility gives some sort of status, while effective contraception foresight and planning are not always easy. This has led to much poverty and social deprivation.

But a number of factors ameliorate this deprivation. The children support their families by working on farms and in urban jobs. In fact, this demand for family labour applies to all faiths. The poor will always tend to overbreed where they fear that few children will survive into adulthood. Even where there was little mechanical restriction on fertility as in Goa and southern India, the Hindu birth rate was higher than the Roman Catholic, almost certainly because of non-religious factors of caste and occupation (Roy et al. 1991, 353–8). The decline in the number of children born to followers of particular religions may correlate with a decline in religious observance, but is more likely due to economic advance from some sort of subsistence economy to an urbanized environment where more job opportunities are available to women, particularly where there is part-time work as societies become more service oriented.

Some Hindus and Buddhists have opposed contraception because they believe it interferes with reincarnation. Sri Lankan Buddhists see it as against the order of nature and that therefore it is better to follow natural methods. The prevention of birth is equated with the prevention of rebirth of someone who might otherwise have gone a stage further towards ultimate release. On the other hand, there is no objection to abstaining from sexual intercourse. Indeed, these faiths encourage such abstention as part of a progression away from being controlled by desire.

The main conclusion is probably that although opposition to contraception contributes to accelerated population growth in some faiths, such growth is even more strongly encouraged by local socio economic factors which determine optimal family size combined with the elimination of mass diseases other than AIDS. These factors, however, are themselves conditioned by the underlying religious ethos which includes such non-literate issues as concern for ancestors and the perpetuation of a family line.

10.7 Violent religious practices reducing pressure on the environment

Just as there are religious practices which enhance the quality of life and enable people to live longer, so there are others which shorten or even terminate the lives of those who carry them out. The snake-handling (Alther 1975, 532–5) and poison drinking (Schwarz 1960, 405–29) cults of North Carolina have had casualties but only in small numbers spread over long periods of time. Thirty deaths from ritual snake bite in 50 years would have little effect on the population of most villages.

The mass suicide of Jews at Masada and in York, the murder of children by their parents at Jonestown in Guyana, and the mass killings in western Uganda, obliterated whole communities. The Skopsi in Russia castrated themselves and thus ensured their extinction as a sect. But these are relatively isolated incidents. The effects on the environment are likely to have been transitory. The neighbourhoods involved soon reverted to the environmental state which had existed before these tragedies.

More extreme examples have left enduring demographic traces. Examples are the human sacrifices which occurred in a wide range of pre-colonial South American societies. This became more frequent with the mass killing of war captives in the century before the Spanish conquest. The religion of the Nahua civilization in Mexico centred around human sacrifice. It raided outside its boundaries to obtain victims (Cook 1946, 81–102), sacrificing an average of 700 war captives each year. The total annual killing between AD 1442 and 1507 may have been 12 000.

When Cortés reached Montezuma's capital, he ordered a skull count of victims at the main temples. This came to 136 000 excluding two towers of skulls which were not counted (Tapia 1866). There may have been a mean annual sacrifice rate of 15 000 in a total population of about 2 million so that these regular holocausts were a major factor in controlling the population and balancing it with the available economic resources (Cook 1946, 81–102). In the eyes of the Mexicans these sacrifices ensured the benevolence of the gods who continued to provide a reasonable living for the remaining population. Since the war captives would have been men in their reproductive prime, the long-term loss could have been considerable.

Some suicides have occurred in a war situation. Masada is a treeless tableland whose only water comes from cisterns catching the occasional rain that falls on its surface. In AD 73, its garrison was the last remnant of Jewish rule in Palestine and it took a Roman army of almost 15 000, fighting a defending force of less than 1 000 including women and children, to subdue the fortress. When the situation was hopeless, the defendants, led by Eleazar ben Jair, took their own lives. Only two women and five children, who had hidden in a water conduit, survived to tell the tale. The site was soon reoccupied by lizards and birds, in recent years becoming a mecca for tourists.

Direct killing is not the only consequence of religious wars. A much larger number of non-combatants die because of the disruption to daily life, especially of the agricultural work during the seasons when fighting was most likely to have taken place. The shortage of men as a result of war casualties and the absence in medieval societies of alternatives to marriage could have destabilized the environment. The possible breakdown of the family was at least reduced by the large numbers of women who entered religious life. It has been calculated that in fourteenth-century Frankfurt 6 per cent of adult women were in religious houses (Flick 1930). This may in part be due to the excess of women over men in the populations of late medieval cities. Nuremberg had 1207 per 1000 men, Basel 1246 and Rostock 1295 (Heer 1962).

The conflict between Catholic and Huguenot in France has had profound environmental effects. The Massacre of St Bartholomew's Night in 1572 probably killed 120 000 Huguenot men, women and children as calculated from the fees paid to those who buried the bodies found floating in the Seine. Henry IV granted toleration to Huguenots in the edict of Nantes in 1578, but this was revoked by Louis XIV in 1685, depriving them of all religious and civil liberties. Within a few years more than 400 000 emigrated to England, Prussia, Holland and America, enriching those countries with their skills and impoverishing France, especially its western and southern parts, where they had been numerous.

Religious wars fought in largely uninhabited areas have not had the same environmental effects as those fought over cultivated land. Their main effect has been demographic in the wholesale killing of young men, as in the recent war between Sunni Iraq and Shi'ite Iran. Where the population is very sparse wars can have great demographic effects. Tribal wars in the pre-colonial western Sahara between the Tuareg tribes and other peoples characteristically ended with the complete slaughter of one side, reducing land use for many years.

The Holocaust killed an estimated $6\frac{1}{2}$ million Jews who were overwhelmingly urban dwellers (*Encyclopedia Judaica* 1971, 8.889). The virtual obliteration of Jewish ghettos and even of predominantly Jewish neighbourhoods left many abandoned buildings and cemeteries and drastically reduced urban population densities. The town of Krakow, for instance, had a large Jewish population in 1939; 55 000 from the city perished at Auschwitz-Birkenau and only a few hundred live in the city today. The situation in Warsaw was even worse. In 1900 the city had the largest concentration of Jews in the world, making up almost half of its population. The Germans established a ghetto surrounded by a high wall during the Second World War. Disease, starvation and overcrowding caused thousands to die even before deportation to the Nazi death camps in which in 1942 alone 312 000 were gassed. A further 60 000 died in the ghetto uprising of April 1943. Only a few hundred Jews live there today. The long-term environmental effect is seen in the destruction and postwar rebuilding of both cities, the blight on the land in and around the death camps and the expanded Jewish settlement in the state of Israel. Non-Jewish populations have taken over the vacated areas.

While we react in horror at the systematic killings of Jews by the Nazis, such obliteration of religious communities has occurred a number of times in history. During the First World War, Armenian support for Russia caused the Turkish government to order the deportation of 1 750 000 Armenians to Syria and Mesopotamia. In the course of this forced exodus, about 600 000 died of starvation or were killed by Turkish police and soldiers while en route. Hundreds of thousands more were forced into exile. This depopulated considerable areas of eastern Asia Minor and caused Armenians to make first temporary and then permanent settlements in cities in Iraq, Syria and elsewhere.

Other wars between communities of different faiths have resulted in exchanges of population. The different methods of land use practised by the exchanged populations cause environmental changes. The victory of the Moslem Turks over the Orthodox Christian Greeks in Asia Minor in the 1920s resulted in the expulsion of Greeks from Asia Minor and Moslems from Crete. Much the same has occurred more recently in Cyprus where after a decade of civil strife and increasing segregation of the two populations, the Turkish invasion in 1974 partitioned the country into northern Moslem Turkish Cypriot and southern Christian Greek Cypriot states. Since more people moved south than north there are still many abandoned farms in the north. The south has developed more rapidly into an important tourist destination.

11
The Impact of Religion on the Treatment of Animals

11.1 The direct and indirect effects of religion on animal life

The religious protection of animal life has had a very long history. Since Hinduism and Buddhism believe in the transmigration of souls, their followers have for centuries protected and freed animals which they believe may be the repository of human souls who have lost their way in their search to detach themselves from worldly desires. This has preserved certain species which might otherwise have disappeared.

The evidence comes particularly from Buddhist societies in Asia. In the late twelfth century King Kirthi Nissankha Malla of Polonnaruwa, Sri Lanka, had an edict engraved on an upright stone slab at the great Buddhist stupa at Anuradhapura. It ordered by beat of drum that no animals should be killed within a radius of seven 'gau'. It also gave security to the fish in the 12 great tanks. He backed up his laws by giving people gold, cloth and whatever other kinds of wealth they wished in order to persuade them to obey (Keuneman 1992, 312). One result is that the meditation caves of Buddhism in China and India are likely to be the living and nesting sites of bats and birds, as are the belfries of many Christian churches.

Some Buddhist monasteries near to Bangkok protect large numbers of monkeys. These have such a privileged position that they adopt quite aggressive tactics in getting food from visitors, as do the deer in the Buddhist temples of Nara, Japan (Figure 2.5) which are even located in the central reservations of the city's main roads. Odoric of Pordenone visited a Buddhist abbey near Hangchow in about AD 1330. He described a monk feeding monkeys:

> he took two great buckets full of scraps from the table, and opening the door of a certain shrubbery, which was there, we went therein. Now in this shrubbery there is a little hill covered with pleasant trees and all full of grottoes. And as we stood there he took a gong, and began to beat upon it, and at the sound a multitude of animals of diverse kinds began to come down from the hill, such as apes, monkeys, and many other animals having faces like men, to the number of three thousand, and took up their places round him in regular ranks. And when they were thus ranged about him, he put down the vessels before them and fed them as fast as he was able. And when they had been fed he began again to beat the gong and all returned to their retreats.
>
> So I, laughing heartily, began to say, 'Tell me, prithee, what this meaneth?' And he answered, 'these animals be the souls of gentlemen, which we feed in this fashion for the love of God'. 'But', quoth I, 'no souls be these, but brute beasts of sundry kinds.' And he said, 'No, forsooth, they be nought else but the souls of gentlemen. For if a man be noble his soul entereth the form of some one of these noble animals; but the souls of boors enter the forms of baser animals and dwell therein!' And say what I list against it, nought else would he believe.
>
> (Yule 1913–15, II, 203)

Joseph Needham points out (1954, 190) that an exactly similar story had been told 16 centuries earlier by the ancient Greek historian Megasthenes (c. 350–290 BC) which is important evidence of the long influence of such religious beliefs on these monastic and temple environments.

The Chinese treated the grounds around the tombs of their sacred emperors as parks in which all living things partook of the holy character of the spirits of the dead, thus serving religious and recreational purposes at the same time (Schafer 1962, 280–1). The area in which Confucius and his descendants are buried remains today a natural area. Only the stone tomb monuments have altered the light forest environment.

Even if there is no specific prohibition of killing animals, respect for religious areas would inhibit such action. Christian churches discourage pigeons and bats by fitting special doors and wire mesh rather than killing them. Some mammals now favour churches and churchyards as a habitat. Out of 285 church porches in Northamptonshire, 87 were occupied by bats of four different species (Greenoak 1985).

Fauna will invade areas where religions encourage it. The holiest parts of some Hindu temples are only entered by a limited number of priests, but contain bats and birds. Some temples not only protect monkeys but other animals as well (Figure 2.8). The temple dedicated to the goddess Karni Ma in Bikanir, India, is estimated to be the home of 100 000 rats and it is suggested that the practice of worshipping and feeding them there has gone on for several centuries (Reynolds and Tanner 1983, 227). Here again there is no boundary between these rats and the outside world; they come and go as they please. The rats and monkeys protected in Hindu temples come and go over the borders of the sacred areas refreshing themselves and being refreshed from outside.

The sacred area surrounding the pilgrimage centre of Katagarama in southern Sri Lanka covers about five square miles (12 sq. km). It contains Buddhist, Hindu and Moslem shrines, with monkeys, birds and lizards in abundance living under protected conditions. These animals gain their livelihoods from the rubbish left behind by the thousands of pilgrims as well as from religion-inspired benevolence.

People have always raised animals for food and fur but the large-scale industrialization of the processes is largely a modern product of the Christian West. It has brought serious challenges to all religions over the treatment of animals. There is a tension between the pressures to achieve an even more efficient production of low-cost food and the growing concern for the price it demands in animal suffering. Although some restraints do exist, there have been some cruel distortions of animals' lives. Dairy production demands methods much to increase lactation and to rob the cow of her calf at an unnaturally early stage. Pork and bacon producers produce grossly large pigs which have proved to be chronic invalids suffering from arthritis and crossed eyes (Midgely 1992, 28). Chickens spend their lives in tiny enclosures in vast batteries to increase and facilitate egg production. Much land is also devoted to hunting. This includes fox coverts, and both woodland and moorland for raising deer and game birds.

Such animal production also has profound effects on land and water use. It takes something like ten times as much land and many times as much water to obtain the same amount of protein from animal than from plant foods. This has greatly distorted land use patterns, turning much potentially arable land and forest into pasture, and where irrigation is needed, diverting much water. The accumulation of large amounts of dung in crowded cattle feeding lots poses serious problems of disposal. It is due to some extent to a religion-inspired conscience that public opinion over these problems is now moving quite fast in a constructive direction.

It is in Christian lands that the main pressure for improving the care and protection of animals has come. The World Wildlife Fund is the consequence. There is also a proliferation of organizations at national level. In Britain this includes the Royal Society for the Prevention of Cruelty to Animals (RSPCA), the People's Dispensary for Sick Animals (PDSA) and the Royal Society for the Protection of Birds (RSPB). There are also many organizations rescuing unwanted pets, giving them homes and providing cemeteries when they die. National and international legislation now protects natural habitats and endangered species such as whales, elephants and tigers. The whole continent of Antarctica enjoys such protection.

11.2 The effects of religion on particular animals

The origins of the Hindu veneration of the cow are ancient and unknown. It may have originated in the same idea of substitutionary sacrifice that applied to cattle in ancient Israel. It may also have grown out of its utility for farmers as an essential draught animal which also provided meat, hides, milk, manure and fuel. It probably grew in importance with urbanization, the shortage of fuel and the ritualization of the cow by a succession of historic charismatic figures. More recently Mahatma Gandhi frequently drew attention to the moral as much as ritual importance of the cow to contemporary Indians. He called the cow a companion and its veneration the distinguishing mark of the Hindu:

> the cow was in India the best companion. She was the giver of plenty. Not only did she give milk but she also made agriculture possible (1954, 3).
>
> Cow protection is the dearest possession of the Hindu heart. It is the one concrete belief common to all Hindus. he who does not believe in cow protection cannot possibly be a Hindu.
>
> (Manley and Neech 1950, 111–12)

The pro-cow movement was also politically useful in the wider campaigns against the beef-eating Moslems and British.

Hindu cow worship, however, goes further than protection. To eat its five products is the most powerful means of purification. The religious importance of the cow will probably continue as it provides a popular ritual focus and must remain a centre of a network of economic usages. But it is unlikely that the originators of the worship anticipated the problems it would cause.

This view of the cow appears anachronistic and ignorant of effective animal husbandry. To Western eyes it presents a ridiculous picture of cows wandering the main streets of many Indian towns so that other aspects of the scene tend to be forgotten (Figure 2.6). Besides milk, even cow dung and urine have religious uses in household shrines. Various Hindi phrases in current use speak of the cow as being like a mother and as a channel to heaven, giving it a uniqueness without a parallel in Western languages. On the other hand, this veneration is seldom accompanied by the high standards of animal husbandry required in modern Western farming. In India cow killing even by accident has caused communal rioting. Devout Brahmans and caste Hindus perform cow-worshipping ceremonies at funerals and at festivals relating to Krishna and Lakshmi.

Reverence for the cow has had important social and economic consequences. The Indian government had an Expert Committee on the Prevention of the Slaughter of Cattle which estimated that there were 20 million head of uneconomic cattle in the country (Nandra et al. 1955), although this is thought to be an exaggeration. Economic uselessness is a matter of degree. Even a decrepit animal will produce dung to be used as household fuel and for improving the soil even if its meat will only be eaten by lower sub-castes. It has been estimated that the amount of dung used as fuel annually in India is the equivalent to between 35 and 131 million tons of coal or 68 million tons of firewood (Parrack 1969, 38). This is one argument for preserving the cow population since even the elderly cow contributes dung.

The cow is a more costly milk producer than the water buffalo. The latter are important for ploughing as there are not enough cows to provide a pair of bullocks for as many as one-third of all small farmers (Harris 1966, 53). This is partly because the demand comes at the peak season for ploughing (Shastri 1960, 1592). On the whole, Hindus tend to keep cows and Moslems water buffaloes.

A study in the holy city of Banaras (Lodrick 1981, 241–2) showed that Moslem preference for water buffaloes was because they yielded a higher valued and priced milk. On the other hand Hindus, particularly those of higher caste, who kept more milk-producing animals than the Moslems, preferred cows to buffaloes regardless of the 20 per cent lower cash return. Indian Christians, Moslems and some Hindus eat beef. It is probable that the Outcastes have beef in their diet because they dispose of the carcasses of dead animals.

Hindu farmers are careful in husbanding resources. They calculate carefully what forage to give what animals, and do not feed the whole herd indiscriminately (Srinivas 1958, 4). In one Hindu village the only exploitations were by powerful well-to-do men with too many cows and by the landless people wanting to get into cattle ownership who grazed their beasts on common land (Freed and Freed 1972, 399–408). The concept of cow sanctity often induces farmers to borrow money to replace one which has died from disease, old age or accident. They will do this even though it risks the ever-present danger of defaulting and losing the land.

The veneration of the cow has led to the survival of large numbers of partially useless animals. They are allowed considerable public licence, wandering the city streets and eating food exposed in the shops and markets as any tourist can see in Calcutta. In the crowded Gangetic plain this has led to a chronic shortage of fodder and the deterioration of the breed (Mukerjee 1938, 142–3). Cows abandoned by their owners are also a costly nuisance in the countryside where they break into fields of growing crops (Mayadas 1954, 29).

There is perforce a scale of utility against which the preservation of the useless cow is measured. When possible starvation and deprivation face the owner of cows, he will inevitably sacrifice animals not essential for his survival (Dandekar 1959, 352). Even these will go for slaughter if the pressure is extreme enough (Parrack 1969, 36–8). The 1943 Bengal famine coincided with an increased demand for beef from the armed forces fighting Japan. There was a rush to sell cattle for slaughter. This led to the legal prohibition of killing cattle needed for ploughing, milk production and the breeding of bullocks (Knight 1954, 141). But in general Hindu cattle owners will neglect their animals to death rather than slaughter them and only sell them under duress directly or indirectly to butchers (Roy 1955, 15).

Where the cow has economic uses these are subordinated to religious considerations. Certainly cows are slaughtered but those who are involved in this are both socially and religiously segregated from mainstream Hindu society. There are sanctions even against the very reasonable need to shoo cows out of your shop. Although cow dung is always useful, religious and social norms prevent the full utilization of the cow beyond this. When it dies the carcass is removed by Outcastes or just thrown into the river.

Hinduism is unique in having large numbers of homes (called *gowshalas, pinjrapoles* or *gosadans*) for cows. These are complex institutions maintained by contributions from the devout, somewhat analogous to the dog and horse homes supported by Westerners. Recent data show that in India they accommodate some 550 000 cows of whom 174 000 are kept there until they die. Therefore they contribute nothing to their own maintenance other than the residual value of their carcasses, hides and dung (Lodrick 1981, 201). The existence of these homes has been recognized in the past by the secular government of India. Under political as well as religious pressure its First Five Year Plan anticipated that there would be an increase of 160 homes which would serve a further 320 000 cows. There has been a steady increase in the numbers of these institutions over the years.

It is estimated that each animal in these homes eats 10 kg of forage per day, amounting to an annual consumption of 1.66 million kg of fodder in a country facing chronic as well as seasonal shortages of fodder. They compete directly with useful cattle for available supplies in times of scarcity and the administrators of these homes with their greater purchasing power can effectively outbid local farmers. The use of agricultural land to produce fodder for useless animals because of religious beliefs is an important environmental issue.

On the other hand during the 1972–74 Gujarat drought these homes were able to keep alive some useful animals placed there by their farmer-owners who could no longer maintain them. Although many died, some were returned to the agricultural sector when the rains finally came. Some farmers pay rent to these institutions on the understanding that if their cows begin to provide milk, they can be reclaimed (Gourou 1963, 125).

Cattle worship affecting the environment is not confined to India. The Nilotic tribes of the southern Sudan, notably the Dinka, Nuer and the Shilluk and also the Masai of eastern Africa, depend on cattle for their livelihoods in the extensive grassy plains of their native habitats. They have been described as 'cattle cultures' (Herskovitz 1926). Cattle keeping is the connecting link between many aspects of their social, economic and political, as well as their religious, lives. They use them in rituals of male puberty, and occasionally sacrifice them. The Dinka reverence their cattle and call them 'brothers'. They use every part of their bodies. They divide up the meat into shares which reflect the recipients' relationship to the owner, and use the hides for clothing and the dung for fuel. They sometimes mix meat and milk with blood as food. On reaching puberty each youth goes through ceremonies at which he receives a totem animal dedicated to his ancestors, called his 'song bull' which is castrated on his behalf to certify his arrival at manhood. The environmental effect of the quasi-religious relation between these cattle-keeping tribes and their herds is to strengthen their pastoral lifestyle and make cattle the central feature of their economies.

The economic life of the non-Hindu Todas of the Nilgiri hills in southern India centres on their sacred buffaloes (Mandelbaum 1941, 22–3), and the Chins of north-western Myanmar hold the semi-domesticated mithun (wild ox) to be sacred, linking its use in sacrifices to fertility and the spirits (Simoons 1968, 260). The Sukuma of Tanzania dedicate goats and cows to their ancestors. The milk of the latter can only be drunk by male relations on the father's side. They are kept alive and given distinctive bells round their necks as a channel for addressing the ancestors who are causing difficulties for certain families. Such correlations between spirits and dedicated animals are widespread.

Hinduism not only protects cows. The non-violent concept of *ahimsa* forbids the taking of the life of any living thing (Manu 6: 45–81) and is an important determinant of wider Hindu behaviour (Brandon 1970, 48). Since it abjures material rewards no Hindu could expect any immediate return from helping to preserve the lives of economically useless cows. Jainism and Buddhism also believe this and see *ahimsa* as leading to meritorious acts which can in the long run contribute to the release of individuals from the cycle of constantly changing but seemingly endless forms of existence.

The reluctance of both Buddhists and Hindus to kill has arguably made them less aggressive than either Moslems or Christians. They protect rabid dogs and disease-carrying insects. Tibetan monks may put their robes out of doors during the night so that the low temperatures will kill the bugs (Ekvall 1964b, 76), but even this action would send the perpetrator to one of the eight cold hells.

Jews, Moslems and some Christians follow the Mosaic code, and many Hindus, Jains and Buddhists are vegetarian. Many people in all faiths avoid eating meat, although their reasons may be philosophical as much as religious. The Orang Asli of Malaysia restrict the eating of animal protein to men (Bolton 1972, 789–99) although this may be at least partly because the men in that society eat first and women get the leftovers. Where meat is scarce this is likely to be very little. Where the animals concerned are no longer available locally, and exist only as a memory on postcards and in folk tales, such taboos become merely symbolic.

11.3 The use of animals in sacrifices

The main uses of animals are for draft and diet, but they are also used in religious sacrifices. We have seen historical examples in tribal religions, in ancient China, Israel and Mexico, and up to the present day on the Moslem Hajj. They have always posed the obvious problem of maintaining flocks. For instance, although some tribes in East Africa and the Sudan practise ritual animal sacrifice, the need to conserve the stock limits the numbers they can use. The uncertainties of livelihood demand a calculating use of their stock and a need to accumulate cattle. The Nuer, for instance, are quite capable of substituting a cucumber for a cow in their worship on the grounds that their *kwoth* (venerated sky spirit) had not paid sufficient attention to their previous sacrificial requests. Cattle can be too economically valuable and prestigious to be used in regular religious ceremonies. They tend to be used for feasts or gifts or in resolving serious crises. But they will eat the carcasses of animals whose death was unforeseen.

Some groups in New Guinea breed pigs for special ceremonial feasts which they hold after they have multiplied. If they did not delay for this the pig population would overrun each community's habitat. Thus the people maintain a balance between pig numbers and the capacity of the environment to sustain them (Rappaport 1967, 17–30). Similarly, the Chukchee tribe of north-eastern Siberia used sacrifices to reduce their caribou herds and provide themselves with protein. They killed male fawns and big bucks pairing each with a female fawn, thus holding down herd accumulation, at the same time preventing sharp status differences developing in the community from a large number of breeding females being retained (Bogoras 1904–9).

Haruspiciation is the method of predicting the future by examining the entrails of specially killed animals or fowls. It is common in some cultures (Murra 1965, 186). The Sukuma of Tanzania mainly use young chickens. The environmental effects are relatively small, deriving only from the small number of creatures sacrificed.

The levitical system of ancient Israel involved the sacrifice of many animals. The purpose was to symbolize the divine way of forgiving sin. It included many sorts of offering, sacrificing a variety of animals from cattle to pigeons depending on the purpose and the wealth of the offerer. The animal sacrifices in some other religions derive from this.

A significant example of a religious sacrifice occurs on the Moslem Hajj. Since every pilgrim kills a sheep in a short time in a small area, the output is vast and concentrated. Even in the late nineteenth century the sacrifices of some 30,000 people had made the large number of carcasses a health hazard. The earlier religious requirement that pilgrims give to the poor what they could not eat themselves became unworkable, and the Meccan authorities had to organize mass burials and the lavish use of lime. The small resident Meccan population which at most contains a few thousand indigents, could not possibly dispose of the sacrifices of millions of pilgrims. There are now centralized efforts to store the surplus meat from perhaps half a million animals and to move it in bulk to countries in which there is a shortage of animal protein. Also, the sacrificial animals use up enormous quantities of imported fodder on the way to Mecca and deprive the areas from which they come of any betterment their products might confer.

Religious compulsions sometimes involve the mass destruction of domestic herds. In 1853 a South African Xhosa Christian convert returned to his home area and began preaching a religious message that involved killing the tribal cattle in return for immediate earthly benefits. His niece Nongqawuse of whom he was guardian, developed this theme stating that her prophecies came from her conversations with spirits. These spirits had told her that the Xhosa dead were preparing to rise again bringing with them wonderful new cattle, but that the living members of the tribe must slaughter their own cattle and destroy their food stocks as both were contaminated and impure. Furthermore, they should not cultivate in the coming season but instead build new cattle enclosures and grain storage pits for the cattle and corn which would appear miraculously. They should also build new houses and wear new ornaments. On top of this the blind would see again, the deaf hear, cripples walk and the old would become young again.

In support of her message was the fact that many saw the lung sickness epidemic of 1855 which had killed at least half of their cattle as a sign that the ancestors were both troubled and wished to communicate with them. With such a generalized belief it was not difficult for a charismatic prophet to develop her message to state that if all impure and evil things disappeared, the Xhosa world would be a better place.

Other factors added impetus. The Xhosa had been in violent conflict with not only the colonial government but also with the local settlers on and off for the preceding half-century. They were being progressively squeezed off their ancestral lands into the territories of other tribal groups to the north. While Xhosa men would have responded to this aspect of the prophet's message, to their women it offered a totally different liberation. They bore the main burden of cultivation and rejoiced at this message that they would get crops without having to work their fields at all (Soga 1857). It was not surprising that they became the most enthusiastic of the prophet's followers.

The consequences were catastrophic. By January 1857 the Xhosa were estimated to have slaughtered 400 000 head of cattle and many did not cultivate at all. The tribe's population dropped from 105 000 to 25 000 between 1857 and 1858. As a result they no longer had enough people to occupy their traditional land and could no longer expect the colonial authorities to recognize all as theirs. The resulting environmental vacuum allowed the expansion of white settlement and the retraction of the area of land which had hitherto been quite adequately cultivated by traditional tribal methods. Peires wrote, 'Their national, cultural and economic integrity, long penetrated and undermined by colonial pressure, finally collapsed. Nongqawuse and Sir George Grey (the Governor) had irrevocably transformed the Xhosa nation into South Africans' (1989, 310–13, 321).

These ideas were a development of existing Xhosa religious beliefs. Like many other Bantu peoples they believed that the dead did not really die in the biological sense but were a continuum with their living descendants, and that they imposed misfortunes on the living. This disaster therefore replaced one religio-cultural utilization of the environment with another. The second was more productive and transformed the Xhosa environment largely for the benefit of other groups. There are some parallels here to the Sun Dance religion of the Sioux in the 1890s who were then at the end of their resistance to white expansion.

12
Religion and Environmental Management

12.1 Secular authorities and religious activities

People generally view environmental management as a political activity with little stress on religious considerations, even in states which are officially religious such as Israel, Saudi Arabia and Pakistan. But all generally recognize that the provision of housing, roads, hospitals, water, sewage and food have a religious dimension. Land use planners, whatever their faith, will be concerned with quasi-religious issues which range from the conservation of religious buildings and cemeteries and the prevention of disasters on mass pilgrimages to minor problems such as the widening of a road by a temple or tomb or the provision of safe water used for sacred purposes.

Crowd control in pilgrimage sites also requires careful planning to prevent panic and the crushing to death of large numbers of people. In such situations a single accident will block the flow of traffic. The police must not only control crowds of people whose languages they may not know, but check both the inevitable criminals who come to live off the pilgrims and the possible appearance of religious terrorists. This is an ever-present possibility in Israel and Islamic countries. The deployment of armed police on sacred sites is now commonplace.

Crowd sanitation at religious events also demands detailed planning. The problem increases where the crowd is composed of people from different cultures with different hygienic practices, who may have religious and cultural objections to the use of public lavatories in the direction they face and their lack of privacy. During the Hajj month the waste from 2 million pilgrims is deposited in a desert area without any access to running water.

It is necessary to manage the medical care of the large numbers of pilgrims at Lourdes, Mecca and Banaras and to make arrangements for those who die there. In any such crowds there will be inevitably a number of medical emergencies and in a million pilgrims on the Hajj there may be a daily total of 1000 people needing medical attention (Hodgetts and Cooke 1999, 957). There is also a need to prevent epidemics for which there will be a heightened risk when, for instance, the Hajj coincides with the hottest months of the year.

In religiously divided societies believers can be volatile. Minor actual or rumoured events can spark violence. The demolition of the mosque at Ayodha by Hindus sparked off riots across India for which secular authorities had to have emergency plans to deal with arson, injuries and murder. However, except for some fundamentalist groups in Islam and Hinduism, the leaders of all religions and their institutions support the international organizations working for world peace. They also combine with secular powers to support United Nations peacekeeping forces and international welfare organizations which reduce the effects of inter-religious violence. While such activities may not produce enduring peace, they have on occasion prevented armed confrontations and saved the lives of thousands. Nevertheless, religious terrorism continues, involving bombs, bullets and even suicide missions which kill or maim indiscriminately. This detracts from the possibilities of rapprochement and raises the costs of security astronomically.

The disposal of the dead can be a major management issue. Only in desert environments do they not occupy land usable by the living. The graves of the illustrious dead now provide homes for the homeless urban poor of Cairo. Local regulations limit the size of cemeteries where the price of burial cannot meet the value of land needed for housing and industry. Shanghai now prohibits the interment of the dead and makes special arrangement in a specially decorated ship for the ashes to be put into the sea outside the city boundaries.

12.2 Current religious activities and environmental management

Religions are concerned with global issues such as deforestation, soil erosion, urban pollution, global warming and the depletion of the ozone layer. But their role in the management of such issues can only be indirect. To have a policy for the environment that is more than a neighbourhood scheme, it is necessary to have a central authority which is responsive to both secular and religious concerns.

In tribal religions environmental ideas are likely to be limited to the social horizons of families or villages concerned with the rhythms of life: birth, puberty, marriage, child rearing and death. They can involve the preservation of springs and sacred groves. The boundaries of one culture and its religious practices merge into those of its neighbours. But the people can destroy both themselves and the environment which supports them. The Homokam Indians of New Mexico created a long and complicated canal system for agriculture over 1000 miles (1600 km) in length. But this did not stop their social extinction because their expanding population deforested their area and killed off the game. Their religion failed to prevent their disappearance from an environment which they destroyed and which then progressively destroyed them.

Eastern religions with their theologies of an infinite number of transmigrated existences do not encourage theologically based endeavours for the environment because they do not see their next incarnations as being a significant part of possible billions to come. Hinduism, in particular, lacks structures which would enable it to manage the environment. Its caste system and decentralization tend to confine all initiatives to the level of the locality or the small group. It required a charismatic individual Hindu such as Mahatma Gandhi to influence others towards secular legislation for the protection of cows. Hindus may dominate an area but they cannot, because of the nature of the caste system, speak with one voice on any issue except perhaps their hostility to Moslems. Also Hindu sacred literature is written in Sanskrit, a dead language, and speaks in such abstract terms as to be of little value for the creation of an environmental policy.

Buddhism in Myanmar used to have a national monastic structure but this was abolished in the early days of the British occupation. Such a structure has only a nominal existence in Thailand and never existed in China. The Dalai and Panchen Lamas in Tibet combined religious with secular authority. But Buddhist authorities have little concern with the environment because of the faith's doctrines of detachment from earthly concerns and its immaterial philosophy. They could not logically develop an intellectual commitment to such a material issue as the protection of the environment. The only possibility would be an elaboration of doctrine such as that which treats trees as possessing the entity which is capable of transmigrating (Figure 6.5). The written doctrines of the Buddha are too esoteric to be used for public enlightenment, especially as they are written in Pali, a language which is no longer spoken.

Christianity combines an acceptance of the command to people to 'multiply and replenish the earth' with a balancing injunction to act as stewards of the environment. It also generally recognizes the biblically inspired separation of church and state. Church officials are stewards of their land and buildings, and manage pilgrimages, burials and sacred sites. But outside these clearly religious areas, secular authorities are responsible for the environment. The Roman Catholic Church and many national churches have an institutional structure which can make broad recommendations, if not decisions, affecting the wider environment. Smaller groups such as the Benedictines, Amish and Hutterites practise almost a theocratic concept in managing their own lands.

Islam's identification of political with religious authority gives national governments of Moslem countries this dual power of decision over the environment. The basic practices for the individual are the Five Pillars, with the Hajj pilgrimage having a unifying function for all believers. The opinions of individual theologians beyond these Five Pillars can be ineffective unless sustained by the Islamic state. None have a specifically Islamic policy for the environment. Indeed, Moslem Iraq created an environmental disaster in Kuwait by setting fire to all its oil wells before its forces retired.

Religious charities, especially in Christian countries, contribute to the development of agriculture, hospitals and schools. They continue to play an important role in the preservation of populations which have suffered disasters. They also support those neglected by secular society. In the developed world this function has been largely taken over by secular welfare systems, but in developing countries where the state is unable to take on any additional expenditure, religious groups step in. Their charities carry out substantial health support functions as well as providing support for work beyond the capacity of the state, such as returning lost individuals to their homes and the long-term care of the abandoned and handicapped.

12.3 Future concern of religions for the environment

All religions in principle include the ultimate management of the environment as one of their objectives. In the monotheistic faiths of Judaism, Christianity and Islam this is seen as a divine activity in a remade earth after the demise of our present material world. In Buddhism and Hinduism it is the ultimate culmination of the process of birth, death and rebirth.

But the present world is primarily concerned with the management of the earth before its eschatological transformation. Its stewardship is implicit in the Old and New Testaments and the Koran. The Jews had rules for many issues, notably agricultural husbandry, slavery, land fallowing, stock management and the safety and hygiene of buildings. Jesus used many examples of such activities in his parables. Although Mohammed's emphasis was more urban and mercantile than agricultural, it is clear from the Koranic stress on generosity and almsgiving that he also associated good resource stewardship with true religion.

But believers have often failed fully to follow out such concepts of stewardship. Even as little as a century ago the risks were far less obvious than they are today. In the past Judaism, Christianity and Islam have emphasized the welfare of individuals, families and communities but were less concerned with the impacts of land and weather outside their daily horizon. Likewise, their time frame was short and they had no way of measuring long-term trends. It is only in the past century that we have had worldwide climatic measurements and only in the past half-century that these have covered all climatic variables from a dense network of points in 'real time'.

Those who believe in a spirit world pay more attention to the sites associated with them. For example, Burmese villagers would make offerings to a tree in which the locally powerful spirit resides in order to protect it from damage, but they would be less interested in protecting other trees in the nearby forest which do not have spirits in their branches. The contemporary Ifugao rice cultivator in the Philippines placates the spirits related to his terraced fields, but not to the spirits in fields belonging to his neighbours.

There are other limitations to the application of religion to environmental management. Religious leadership is often unable to inspire or motivate where no economic gain is apparent. A concept of other worldly reality can easily be submerged in concern for non-religious practicalities such as computer processing, money transactions, the details of a sewage farm, a factory conveyor belt or a stockpile of ammunition.

There is also the question of scale. People are more able to relate to local than global needs. Communities will preserve wells and their surroundings but give less support for a waste purification plant serving millions. This local focus also applies to erosion and pollution. Most religious groups would be slow to change long-established practices to benefit people in other continents or even in another part of the same country.

13
Religion and the Environment: an Overview

Most religions view the natural environment as a harmonious unity, conceived by an external and transcendent mind. They further teach that this applies not only to the world as a whole but also to the neighbourhood and personal level which people experience. As a corollary they recognize that religious groups have a responsibility to preserve, and where possible to enhance, the environment so that humans in general can survive and prosper.

Isolated religious communities have usually succeeded in maintaining a balance in their environments between immediate and future needs. This has often been a condition of survival. As they have expanded in numbers and covered wider areas this has become more difficult. The focus has widened and has evoked a lowered sense of personal responsibility.

There are a few religious communities which have always placed environmental responsibility high on their agenda but these are small. They include some Roman Catholic orders as well as Protestant groups such as the Mennonites. The great majority of Christians belong to larger groups where environmentally friendly lifestyle has only become a live issue in the past half-century. These have a greater and more complex impact.

The Hindu concept of reincarnation, even when it involves the transmigration of souls, is vague and lacking in self-sacrificing dynamic. Therefore, it offers little incentive for environmental concern. The concept of Brahma requires the individual ultimately to be absorbed into an amorphous supermundane power which has no characteristics. Every individual owes his identity to his caste or sub-caste. This largely determines his outlook since there are no religious institutions which bind together castes into a common viewpoint. Different beliefs add to this fissiparous tendency. The many mythical gods have only such functions as communities and sub-castes have incorporated into their rituals. Thus each group creates and uses its own environment within broad and ever-changing religious requirements in which only the veneration of the cow has constant environmental relevance. The conservation of insects and animals such as temple monkeys and snakes has been too sporadic and idiosyncratic to have had anything more than an extremely local effect. Jainism has much the same shortcomings of belief but with the added ideal of self-starving in order to avoid killing other living creatures. Because of the small number of Jains, this can have little environmental significance except to encourage the proliferation of animal life and to reduce the believing and practising population towards extinction.

Buddhism questions the existence of a detached conscious deity and denies the reality of the natural world. It believes in the abolition of personal identity causing individual nullity at the end of innumerable soulless rebirths. It even sees the acceptance of faith as a failing. It is difficult to see how a Buddhist can have a concept of the enduring environment without violating the basic principles of the faith. But in practice Buddhists have turned the founder's philosophy into a religion which assumes something like divine power. The behavioural guidelines provided by monks who have no social message and no social obligations except to themselves, threaten all local activities, branding them as failings which can only delay the attainment of extinction. The result is that the faith, from its own postulates, undermines any tendency towards environmental concern.

Hinduism and Buddhism therefore suffer from a paradigm of functional hopelessness. There is nothing in their basic doctrines to lighten the ongoing burdens of people whose existence can never be guaranteed for more than a year between harvests. As a response to the resulting hopelessness, individuals, while retaining the label of faith, have created local religions with multiple gods, providing at least the hope of some immediate relief. Thus there are certain communities, now as in the past, which have reached a more positive relationship between their religious practices and the environment on which they depend for a livelihood. This is notable, for instance, in the hydraulic societies of Sri Lanka, Cambodia and Bali.

The Chinese and Japanese religions with allegedly millions of gods similarly lack any institutional centralization which would guide environmental impact, especially since some gods may be mutually antagonistic. Where the Chinese authorities have shown concern for the environment as in controlling the cutting down of forests, the impetus has been secular rather than religious.

In general, the multiplicity of Eastern faiths, especially with their localized forms of ancestor worship, has tended to restrict the influence of each to a small area. The exception has been where a powerful religiously minded ruler with an efficient administration or a strongly distinctive group have had a wider effect. For example, Hindu religious structures in Bali have overarched social and political boundaries.

The three great monotheistic faiths each have a relatively well-defined world view and eschatological expectation. They see nature as having an external divine causation and being subject to discoverable laws. This has made their approach detached and rational. It has generated a closer relationship between their practices and the environment. In practice, however, they find complex tensions between conservation and exploitation and between environmentally aware, usually rural, communities who derive their livelihood directly from the land and sea, and the less aware, usually urban, populations whose dependence on industrialized farming and non renewable resource extractions risk making them less aware of the threats these pose to human life in the long run. There is a need from a religious point of view to emphasize the need for radical changes in the lifestyle of both individuals and communities.

The environmental influence of Judaism today is most clearly seen in its settlement patterns in Palestine and in the dynamic influence Jewish intellectuals have had on the thinking of other faiths. Christianity is important in being the most widely universal of all faiths, in influencing lifestyle and land uses on a global scale and paradoxically in being the faith of the most environmentally exploitative societies today. But at the same time, those from a Christian background have pioneered the advocacy of global conservation and sustainable development. Islam is second only to Christianity in global spread. It differs in its more strongly enforced dietary and temperance regulations, its partible inheritance laws which make for property subdivision, and in its theologically based passivity and resignation in the face of environmental challenges.

Bibliography

Abruzzi, W. S. (1987) 'Ecological stability and community diversity during Mormon colonization of the Little Colorado River Basin', *Human Ecology*, 15(2), 317–38.

Ackerknecht, E. H. (1971) *Medicine and Ethnology; Selected Essays*, Johns Hopkins University Press, Baltimore.

Adams, R. Mc. (1966) *The Evolution of Urban Society: Early Mesopotamia and Pre-Hispanic Mexico*, Aldine, Chicago.

Addison, S. T. (1925) *Chinese Ancestor Worship*, Shanghai.

Agrawal, B. C. and Agrawal, S. K. (1972). 'A note on Natra, the so-called "remarriage" among the Hindus of Malwa', *Eastern Anthropologist*, 25(1), 73–80.

Al-Kindi (1966) *The Medieval Formulary*, Levey, M. (trans.) University of Wisconsin Press, Madison.

Alley, K. D. (1994) 'Ganga and gandagi: interpretation of pollution and waste in Benaras', *Ethnology*, 33(2), 127–45.

Altekar, A. S. (1956) *The Position of Women in Hindu Civilisation*, Berarsidar, Benares.

Alther, L. (1975) 'The snake handlers', *New Society*, 34, 532–5.

Ammar, H. (1954) *Growing Up in an Egyptian Village*, Routledge, London.

Amundsen, D. W. and Ferngreen, G. B. (1986) 'The early Christian tradition' in R. L. Numbers and D. W. Amundsen (eds) *Caring and Curing: Health and Medicine in the Western Religious Traditions*, Macmillan – now Palgrave, New York.

Anderson J. N. D. (1950) 'Islam', in J. N. D. Anderson (ed.) *The World's Religions*, Inter-Varsity Fellowship, London, 52–98.

Andrews, E. D. (1963) *The People Called Shakers*, Dover, New York.

Ani, Papyrus of. British Museum MSS 10 470, London.

Arbellot, G. (1970) *Cinq Paroisses du Village xviie–xviiie Siècle*, Études de Demographie Historique.

Arberry, A. J. (1955) *The Koran Interpreted*, Allen and Unwin, London.

Arens, W. (1979) *On the Frontiers of Change: Mto Wa Mbu, Tanzania*, University of Michigan Press, Ann Arbor.

Arrington, L. J. (1958) *Great Basin Kingdom: an Economic History of the Latter Day Saints*, Harvard University Press.

Asso, I de (1790) Historia de la economica politico de Saragossa.

Aubin, H (1966) *Mediaeval Agrarian Society in its Prime: 4. Lands East of the Elbe and German Colonisation Eastwards*, Cambridge Economic History of Europe, Cambridge University Press, Cambridge.

Baer, G. (1962) *A History of Land Ownership in Modern Egypt, 1800–1950*, Oxford University Press, Oxford.

Bahakim, H. M. (1987) 'Muslim parents' perception of and attitude towards cancer' *Annals Tropical Pediatrics*, 7(1), 22–6.

Banerjea, A. C. (1951) 'Note on cholera in the United Provinces (Uttar Pradesh),' *Indian Journal of Medical Research*, 39, 17–40.

Banerjee, A. (1990) 'Comparative curfew: changing dimensions of communal politics in India', in V. Das (ed.) *Mirrors of Violence*, Oxford University Press, Delhi.

Bardet, J-P. (1973) *Enfants Abandonnés et Enfants Assistés à Rouen dans le Seconde Moitié du Xviiie Siècle*, in Hommage à Marcel Reinhard, Societé de Demographie Historique, Paris.

Barrett, D. B. (1968) *Schism and Renewal in Africa*, Oxford University Press, Nairobi.

Barth, F. (1992) 'Towards greater naturalism in conceptualizing societies', in A. Kuper (ed.) *Conceptualizing Society*, Routledge, London.

Bartlett, H. H. (1956) 'Fire, primitive agriculture and grazing in the Tropics', in W. L. Thomas (ed.) *Man's Role in Changing the Face of the Earth*, Chicago University Press, Chicago.

Basham, A. L. (1964) 'Jainism', in R. C. Zaehner (ed.) *Concise Encyclopedia of Living Faiths*, Hutchinson, London.

Baxi, U. (1990) 'Reflections on the Reservations crisis in Gujarat', in V. Das (ed.) *Mirrors of Violence*, Oxford University Press, Delhi, 215–39.

Bayly, S. (1999) *Caste, Society and Politics from the 18th Century to the Modern Age*, Cambridge University Press, Cambridge.

Beardsley, R. K. et al. (1959) *Village Japan*, University of Chicago Press, Chicago.

Bedikian, A. Y. and Thompson, S. E. (1985) 'Saudi community attitudes towards cancer', *Annals Saudi Medicine*, 5, 161–7.

Behrend, H. and U. Luig (1999) *Spirit Possession, Modernity and Power in Africa*, Currey, Oxford.

Bellah, R. N. (1966) 'Religious systems', in E. Z. Vogt and E. M. Albert (eds) *People of Rimrock: a Study of the Values in Five Cultures*, Harvard University Press, Cambridge, Mass., 227–64.

Bellah, R. N. et al. (1985) *Habits of the Heart: Individualism and Commitment in American Life*, University of California Press, Berkeley.

Beloch, K. J. (1937) *Die Volkerungsgeschichte Italiens*, vol. 1, Berlin.

Benedict, P. (1991) *The Huguenot Population of France, 1600–85*, Philadelphia.

Benigni, U. (1911) 'Patron saints', *Catholic Encyclopedia*, Appleton, New York.

Bennett, D. H. (1986) *Interspecies Ethics: Australian Perspectives; a Cross-cultural Study of Attitudes towards Non-human Animal Species*, Environmental Philosophy Series No. 14, Australian National University, Canberra.

Bennett, J. W. (1967) *Hutterian Brethren: the Agricultural Economy and Social Organisation of a Communal People*, Stanford University Press.

Bennett, J. W. (1976) *The Ecological Transition: Cultural Anthropology and Human Adaptation*, Pergamon Press, London.

Bentley-Taylor, D. (1950) 'Buddhism', in J. N. D. Anderson (ed.) *The World's Religions*, Inter-Varsity Fellowship, London, 118–35.

Beresford, M. (1967) *New Towns of the Middle Ages: Town Plantations in England, Wales and Gascony*, Lutterworth Press, London.

Bernadi, B. (1959) *The Mugwe: a Failing Prophet*, London.

Bernard, Saint (1919) 'Sermon in praise of the Templers' in A. Vanderpool (ed.) *La Doctrine Scholastique du Droit de Guerre*, Paris, 204–6.

Bhardwaj, S. M. (1973) *Hindu Places of Pilgrimage in India*, University of California Press, Berkeley.

Bieler, L. (ed.) (1963) *The Irish Penitentials*, Dublin Institute for Advanced Studies, Dublin.

Binsbergen, W. M. J. (1978) 'Explorations in the history and sociology of territorial cults in Zambia', in J. M. Schoffeleers (ed.) *Guardians of the Land: Essays on Central African Territorial Cults*, Mambo Press, Gwelo.
Boardman, E. P. (1952) *Christian Influence upon the Ideology of the Taiping Rebellion*, University Wisconsin Press, Madison.
Bogoras, W. (1904–9) *The Chukchee Jessup North Pacific Expedition*, 7, 372–3, Brill, Leiden.
Boissonnade, P. (1927) *Life and Work in Mediaeval Europe (5th to 15th c)*, E. Power (trans.), Kegan Paul, London.
Bolton, J. M. (1972) 'Food taboos among the Orang Asli in west Malaysia', *American Journal of Clinical Nutrition*, 25, 789–99.
Boniface, Saint (1973) *Letters*, Octagon Press, New York.
Book of the Discipline (1938–42) Pali Text Society, London.
Bownas, C. (1959) 'Shinto' in R. C. Zaehner (ed.) *The Encyclopedia of Living Faiths*, Hutchinson, London.
Boyajian, D. H. (1972) *Armenia: the Case for a Forgotten Genocide*, New York.
Boyer, P. (1994) *The Naturalness of Religious Ideas: a Cognitive Theory of Religion*, California University Press, Berkeley.
Brabo, F. X. (1872) *Inventarios De Los Bienes Hallados A La Expulsion De Los Jesuitas*, Madrid.
Bradfield, R. M. (1971) *The Changing Pattern of Hopi Agriculture*, Royal Anthropological Society Occasional Paper 30, London.
Bradford, J. (1957) *Ancient Landscapes: Studies in Field Archaeology*, Bell, London.
Brandon, S. G. F. (1970) *Dictionary of Comparative Religion*, Weidenfeld & Nicolson, London.
Brass, P. R. (1997) *Theft of An Idol: Text and Context in the Representation of Collective Violence*, Princeton University Press, Princeton.
Braudel, F. (1973) *Capitalism and Material Life, 1400–1800*, Fontana-Collins, London.
Brise-Barre, A. M. (1993) 'The complications involved in attempting sacrifices: the sacrifice of Id Al Kabir Islam in a French suburb'. *Anthropology Today*, 9(1), 9–12.
Brook, S. (1989) *The Club: the Jews in Modern Britain*, Constable, London.
Bucking, B. (1994) 'Japanese religion', in J. Holm (ed.) *Attitudes to Nature*, Pinter, London.
Bush, L. E. (1986) 'The Mormon tradition', in R. L. Numbers, and D. W. Amundsen, (eds) *Caring and Curing: Health and Medicine in the Western Religious Tradition*, Macmillan – now Palgrave, New York.
Bushnak, A. A. (1978) *The Hajj Transportation System*, Hajj Studies No. 1, Croom Helm, London.
Byrd, R. C. (1988) 'Positive therapeutic effects of intercesssory prayer in a coronary care unit population', *Southern Medical* J., 81(7), 826–9.
Cady, J. (1958) *A History of Modern Burma*, Cornell University Press, Ithaca.
Calley, M. J. C. (1965) *God's People: West Indian Pentecostal Sects in England*, Oxford University Press, Oxford.
Capellmann, C. F. N. (1882) 'Pastoral medicine', in D. F. Kelly, *The Emergence of Roman Catholic Medical Ethics in North America*, Pustet, New York.
Caraman, P. (1975) *The Lost Paradise: an Account of the Jesuits in Paraguay 1607–1768*, Sidgwick and Jackson, London.

Carr-Saunders, A. M. (1922) *The Population Problem: a Study in Human Evolution*, Clarendon Press, Oxford.

Carstairs, G. M. (1966) *The Twice-born*, Hogarth Press, London.

Chabot, C. (1827) *Encyclopedia Monastique*, Leroy, Paris.

Chagnon, N. A. (1977) *Yanomamo: the Fierce People*, Holt, Rinehart and Winston, New York.

Chandrasekaran, C. (1952) 'The cultural component of sexual abstinence in Indian life', *Proceedings Third International Conference on Planned Parenthood*, Delhi, 73–9.

Chandrasekhar, S. (1959) *Infant Mortality in India,1901–55*, Allen and Unwin, London.

Chavannes, E. (1910) *Le T'ai Chan; Essai du Monographie d'un Culte Chinois*, Ernest Leroux, Paris.

Chavis, M. and Cavendish, J. C. (1994) 'More evidence on US Catholic Church attendance', *Journal Scientific Study of Religion*, 33(4), 376–81.

Chernela, J. M. (1998) 'Missionary activity and Indian labour in the Upper Rio negro of Brazil, 1680–1980: an historical ecological approach', in W. Balee (ed.) *Advances in Historical Ecology*, Columbia University Press, New York, 313–37.

Chidester, D. and Lidenthal, E. L. (1995) 'Introduction', in D. Chidester, and E. L. Lidenthal (eds) *American Sacred Space*, Indiana University Press, Bloomington, 1–29.

Choudhury, A. R. (1994) 'Hinduism', in J. Holm (ed.) *Attitudes to Nature*, Pinter, London, 53–76.

Coedes, G. (1940) 'Les gîtes d'étape à la fin du xiième siècle', *Bulletin de l'École Française d'Extrême-Orient*, 40, 247–9.

Cole, M. and Scribner, S. (1974) *Culture and Thought*, John Wiley, New York.

Collipp, P. J. (1969) 'The efficacy of prayer; a triple-blind study', *Medical Times*, 97, 201–4.

Comstock, G. W. and Partridge, K. B. (1972) 'Church attendance and health', *Journal of Chronic Disease*, 25, 665–72.

Conybeare, F. C. (1921) *Russian Dissenters*, Harvard University Press, Cambridge, Mass.

Cook, S. F. (1946) 'Human sacrifice and warfare as factors in the demography of pre-Colonial Mexico', *Human Biology*, 18, 81–102.

Cory, H. (1951) *The Ntemi: Traditional Rites of a Sukuma Chief in Tanganyika*, Macmillan – now Palgrave, London.

Cory, H. (1953) *Sukuma Law and Custom*, Oxford University Press, Oxford.

Coulton, C. G. (1925) *Mediaeval Village, Manor and Monastery*, Harper, New York.

Coulton, C. G. (1927). *Five Centuries of Religion: 2. The Friars and the Dead Weight of Tradition, 1200–1400*, Cambridge University Press, Cambridge.

Coyne, W. D. (1948) *Our Lady of Knock*, Catholic Book Publishing Co., New York.

Cragg, K. (1971) *The Event of the Koran*, Allen and Unwin, London.

Crapanzano, V. (1980) *Tuhami: Portrait of a Moroccan*, Chicago University Press, Chicago.

Crawford, M. E. et al. (1989) 'The relationship between religion and mental health/distress', *Review Religious Research*, 31(1), 16–22.

Cribb, R. et al. (1991) 'The mass killings in Bali', in R. Cribb (ed.) *The Indonesian Killings, 1965 6*, Monash University Press, Victoria

Cross, H. E. et al. (1968) 'Cancer of the cervix in an Amish population', *Cancer*, 21, 102–8.

Cross-Upcott, A. R. W. (1956) 'Social aspects of Ngindo bee-keeping', *Journal of the Royal Anthropological Institute*, 86 (2), 81–109.

Cross-Upcott, A. R. W. (1958) 'Ngindo famine subsistence', *Tanganyika Notes and Records*, 50 (June), 1–20.

Cruden, A. (1769) *A Complete Concordance to the Old and New Testament or a Dictionary and Alphabetical Index to the Bible*, Warne, London.

Dandekar, K. (1959) *Demographic Survey of Six Rural Communities*, Asia Publishing House, Bombay.

Daniel, C. (1966) *The Story of the Eyam Plague*, privately printed, Eyam.

Darlington, C. D. (1969) *The Evolution of Man and Society*, Allen and Unwin, London.

Danforth, L. M. (1982) *The Death Rituals of Rural Greece*, Princeton University Press, Princeton, New Jersey.

Deglopper, D. R. (1974) 'Religion and ritual in Lukang', in A. P. Wolf (ed.) *Religion and Ritual in Chinese Society*, Stanford University Press, Stanford, 19–42.

Dickemann, M. (1979) 'Female infanticide, reproductive strategies and social stratification; a preliminary model', in N. A. Chagnon, and I. W. Irons (eds) *Evolutionary Biology and Human Social Behaviour; an Anthropological Perspective*, Duxbury Press, Duxbury. 321–67.

Dobyns, H. F. (1991) 'Do it yourself religion: the diffusion of Folk Catholicism in Mexico's northern frontier, 1821–46', in N. R. Crumline and A. Morinis (eds) *Pilgrimage in Latin America*, Greenwood Press, New York, 53–70.

Dondog, C. R. (1972) 'Lamaism and its influence on the natural growth of the population of the Mongolian Peoples' Republic', *Sov Zdravookhr*, 31, 52–4.

Donkin, R. A. (1973) 'Changes in the early Middle Ages', in H. C. Darby (ed.) *A New Historical Geography of England*, Cambridge University Press, Cambridge, 104–110.

Dopsch, A. (1966) *Agrarian Institutions of the Germanic Kingdoms from 5th to 9th Centuries*, Cambridge Economic History of Europe, Cambridge University Press, Cambridge.

Dorff, E. N. (1986) 'The Jewish tradition', in R. L. Numbers, and D. W. Amundsen, (eds) *Caring and Curing: Health and Medicine in the Western Religious Tradition*, Macmillan – now Palgrave, New York, 8–33.

Dorjahn, V. R. (1958) 'Fertility, polygyny and their interrelation in Temne society', American *Anthropologist*, 60, 838–60.

Douglas, M. (1966) *Purity and Danger*, Routledge and Kegan Paul, London.

Dubois, J. A. (1972) *Hindu Manners, Customs and Ceremonies*, Oxford Unversity Press, London.

Dubos, R. (1973) 'A theology of the earth', in I. G. Barbour (ed.) *Westerners and Environmental Ethics*, Addison-Wesley, Reading, Mass.

Dubos, R. (1974) 'Fransiscan conservation versus Benedictine stewardship', in D. and E. Spring (eds) *Ecology and Religion in History*, Harper and Row, New York, 122–30.

Dugdale-Caley (1817–30) *Monasticon Anglicanum*, W. Dugdale (ed.) History of Fountains, v. 292ff.

Dukes, E. J. (1920) 'Feng-shui', in J. Hastings (ed.) *Encyclopedia of Religion and Ethics*, Clark, Edinburgh.

Dumont, L. (1970) *Homo Hierarchus: the Caste System and its Implications*, Weidenfeld & Nicolson, London.

Dupont-Sommers, A. (1961) *The Essene Writings from Qumran*, Blackwell, Oxford.

Eck, D. L. (1981) 'Tirthas: crossings in sacred geography', *History of Religions*, 20 (4), 336.

Eddy, M. B. (1906) *Science and Health with Key to the Scriptures*, Trustees Will, M. B. G. Eddy, Boston.

Edmonson, M. (1982) *The Ancient Future of the Itza; the Book of Chilam Balam of Tizimin*, University of Texas Press, Austin.

Eickelman, D. F. (1976) *Moroccan Islam: Tradition and Change in a Pilgrimage Center*, University of Texas Press, Austin.

Eickhorn, W. (1959) 'Taoism', in R. C. Zaehner (ed.) *The Concise Encyclopedia of Living Faiths*, Hutchinson, London, 385–401.

Ekvall, R. B. (1964a) 'Peace and war among the Tibetan nomads', *American Anthropologist*, 66(5), 1119–48.

Ekvall, R. B. (1964b) *Religious Observances in Tibet*, Chicago University Press, Chicago.

Elgood, C. (1951) *A Medical History of Persia and the Eastern Caliphate*, Cambridge University Press, Cambridge.

Ellen, R. (1978) *Environment, Subsistence and System*, Cambridge University Press, New York.

Ellison, C. et al. (1989) 'Does religious commitment contribute to individual life satisfaction'? *Social Forces*, 68(1), 100–23.

El Tayeb, E. M. (1976) *Health Problems in the Sudan Associated with the Mecca Pilgrimage*, Dissertation, Ross Institute, London.

Elton, C. S. (1958) *The Ecology of Invasion by Animals and Plants*, Methuen, London.

Elvin, M. (1974) 'The administration of Shanghai, 1905–14' in M. Elvin and G. W. Skinner (eds) *The Chinese City between Two Worlds*, Stanford University Press, 241

Encyclopaedia Judaica (1971), Keter, Jerusalem.

Encyclopaedia Judaica Yearbook, 1990–1. Keter, Jerusalem.

Engineer, A. A. (1988) *Delhi-Meerut Riots; Analysis, Compilation and Documentation*, Ajanta Publications, Delhi.

Enstrom, J. E. (1978) 'Cancer and total mortality among active Mormons', *Cancer*, 42, 1943–51.

Erlanger, P. (1962) *St Bartholomew's Night*, Weidenfeld & Nicolson, London.

Eusebius (1890) *Church History*, Parker, Oxford.

Evans-Pritchard, E. E. (1937) *Witchcraft, Oracles and Magic among the Azande*, Clarendon Press, Oxford.

Evans-Pritchard, E. E. (1939) 'Nuer time reckoning', *Africa*, 12(2), 189–216.

Evans-Pritchard, E. E. (1956) *Nuer Religion*, Oxford University Press, Oxford.

Evers, J. C. (1949) 'The last bison drives of the Blackfoot Indians', *Journal of the Washington Academy of Science*, 39, 355–60.

Farooq M. and Mallam, M. B. (1966) 'Behavioral pattern of social and religious water-contact activities in Egypt – 49 bilharziasis project area', *Bulletin World Health Organisation*, 35, 377–87.

Farriss, N. M. (1995) 'Remembering the future, anticipating the past: history, time and cosmology among the Maya', in D. O. Hughes and T. R. Trautmann (eds)

Time, Histories and Ethnologies, University of Michigan Press, Ann Arbor, 113–30.

Faure, B. (1987) 'Space and place in Chinese religious tradition', *History of Religions*, 26, 337–56.

Fernandez, J. (1982) *Bwiti: an Ethnography of the Religious Imagination in Africa*, Princeton University Press.

Field, M. J. (1960) *Search for Security; an Ethnopsychiatric Study of Rural Ghana*, Faber and Faber, London.

Flick, A. C. (1930) *The Decline of the Medieval Church*, Kegan Paul, London.

Fortier, B. de la (1963) Les enfants trouvés a l'hôpital général de Montreal: 1754–1804, *Laval Medicine*, 34, 442–53; 35, 335–47; 36,351–9.

Fox, R. (1974) *Alexander the Great*, Allen Lane, London.

Francaviglia, R. V. (1978) *The Mormon Landscape: Existence, Creation and Perception of a Unique Image of the American West*, AMS Press, New York.

Frank, J. A. et al. (1985) 'Major impediments to measles education', *American Journal of Diseases of Childhood*, 139, 881–8.

Fraser, G. E. (1988) 'Determinants of ischemic heart disease in Seventh day Adventists; a review', *American Journal of Clinical Nutrition*, 48, 833–6.

Fraser, G. E. et al. (1991) 'Diet and lung cancer in California Seventh day Adventist', *American Journal of Epidemiology*, 133 (7), 683–93.

Freed, S. A. and Freed, R. S. (1972) 'Cattle in a north Indian village', *Ethnology*, 11, 399–408.

Fromageau, G. (1733) *Dictionnaire des Cas de Consciences*, Paris.

Fulop-Miller, R. (1930) *The Power and Secret of the Jesuits*, Owen, London.

Furnivall, J. S. (1956) *Colonial Policy and Practice*, New York University Press, New York.

Gadgil, M. and Vartak, V. D. (1975) 'The sacred groves of India – a plea for continued conservation', *Journal of the Bombay Natural History Society*, 72(2), 314–20.

Gajdusek, D. C. (1973) 'Kuru in the New Guinea Highlands', in J. D. Spillane (ed.) *Tropical Neurology*, Oxford University Press, Oxford, 376–83.

Gandhi, M. K. (1954) *How to Serve the Cow*, Navajivan Publishing House, Ahmedabad.

Ganshof, F. L. and Verhulst, A. (1966) *Mediaeval Agrarian Society in its Prime:1. France, the Low Countries and Western Germany*, Cambridge University Press, Cambridge.

Garner, J. (1974) 'Spontaneous regressions: scientific documentation as a basis for the declaration of miracles', *Canadian Medical Association Journal*, 111, 1254–64.

Geller, S. A. (1984) 'Autopsies', *Mt Sinai Journal of Medicine* (New York), 51(1), 77–81.

Gennep, A. van (1960) *The Rites of Passage*, Routledge and Kegan Paul, London.

Ghosh, S. K. (1971) *Riots: Prevention and Control*, Eastern Law House, Calcutta.

Giles, L. (1987) 'Possession cults on the Swahili coast; a reexamination of theories of marginality', *Africa*, 57(2), 234–58.

Glasse, R. (1967) 'Cannibalism in the kuru region of New Guinea', *Transactions New York Academy of Sciences*, 29, 748–54.

Gordis, R. (1955) *Judaism in the Modern Age*, Farrar, Strauss and Cudahy, New York.

Gorer, G. (1938) *Himalayan Village*, Michael Joseph, London.

Gourou, P. (1963) 'Civilisation et économie pastorale', *L'Homme*, 123–39.

Grader, C. J. (1969) 'Pura Heduwe Karang at Kubitambahan', in J. van Baal et al. (eds) *Bali: Further Studies in Life, Thought and Ritual*, Hoeve, The Hague, 139.

Graham, A. C. (1959) 'Confucianism', in R. C. Zaehner (ed.) *The Encyclopedia of Living Faiths*, Hutchinson, London.

Grant, C. P. (1937) *The Syrian Desert*, Black, Edinburgh.

Gray, R. F. (1963) *The Ssonjo: an Anthropological Study of an Irrigation Based Society*, Oxford University Press, Oxford.

Greenoak, F. (1985) *God's Acre: the Flowers and Animals of the Parish Churchyard*, Orbis, London.

Griscom, J. H. (1857) *Improvements of the Public Health and the Establishment of a Sanitary Police in the City of New York*, C. van Benthuysen, Albany.

Groen, J. J. (1964) 'De levenswijze der Trappisten en haar invloed op het ontstan van coronaie hartziekten en het cholesterolgehalte van het bloed', *Voeding*, 25, 310–13.

Groen, J. J. et al. (1962) 'The influence of nutrition and ways of life on blood cholesterol and the prevalence of hypertension and coronary heart disease among Trappist and Benedictine monks', *American Journal of Clinical Nutrition*, 10, 456–70.

Groussert, R. (1970) *The Empire of the Steppes*, University of New Brunswick Press, New Jersey.

Gutkind, E. A. (1956) 'Our world from the air; conflict and adaptation', in W. L. Thomas (ed.) *Man's Role in Changing the Face of the Earth*, University of Chicago Press, Chicago, 1–44.

Haigh, C. (1993) *The English Reformation: Religion, Politics and Society under the Tudors*, Oxford University Press, London.

Hall, J. R. and Galanter, M. (1989) *Cults: Faith, Healing and Coercion*, Oxford University Press, New York.

Hamarneh, S. (1962) 'Development of hospitals in Islam', *Journal of the History of Medicine*, 36, 366–84.

Hands, H. R. (1968) *Charities and Social Aid in Greece and Rome*, Thames and Hudson, London.

Haney, C. A. (1972) 'Legitimacy, illegitimacy and live birth ratios in a black population', *Journal of Health and Social Behaviour*, 13, 303–10.

Hannay, D. R. (1980) 'Religion and health', *Social Science and Medicine*, 14A, 683–5.

Harrell, D. E. (1975) *All Things Are Possible: the Healing and Charismatic Revival in Modern America*, Indiana University Press, Bloomington.

Harris, I. (1991) 'How environmental is Buddhism', *Religion*, 21, 101–14.

Harris, M. (1966) 'The cultural ecology of India's sacred cattle', *Current Anthropology*, 7(1), 51–65.

Hartley, I. S. (1882) *Memorial of Robert Milham Hartley*, privately printed, Utica, New York.

Heber, R. (1861) *Narrative of a Journey through the Upper Provinces of India*, John Murray, London.

Heer, F. (1962) *The Mediaeval World: Europe 1100–1350*, New Amsterdam Library, New York.

Herskovits, M. J. (1926) 'The cattle complex in East Africa', *American Anthropologist*, 230–72, 361–88, 494–528, 633–64.

Hickey, J. (1984) *Religion and the Northern Ireland Problem*, Barnes and Noble, Totowa, New York.
Hill, R. (1969) 'Islam in the Sudan', in A. J. Arberry (ed.) *Religion in the Middle East*, Cambridge University Press, Cambridge, Vol. 2, 187–202.
Hodgetts, T. J. and Cooke, M. W. (1999) 'The largest mass gathering: medical cover for millennium celebrations need careful planning', *British Medical Journal*, 318, 10 April, 987.
Holdsworth, C. J. (1973) 'The blessings of work; the Cistercian view', in D. Dukes, (ed.) *Sanctity and Secularity: the Church and the World*, Blackwell, Oxford.
Holmberg, A. R. (1952) 'The wells that failed', in E. H. Spicer, (ed.) *Human Problems of Technological Change*, Russell Sage, New York, 113–23.
Horden, P. and Purcel, N. (2000) *The Corrupting Sea: a Study of Mediterranean History*, Blackwell, Oxford.
Hostetler, J. A. (1993) *Amish Society*, Johns Hopkins University Press, Baltimore.
Hostetler, J. A. and Huntington, G. E. (1967) *The Hutterites in North America*, Holt, Rinehart and Winston, New York.
Hudson, E. H. (1963) 'Treponematosis and pilgrimage', *American Journal of Medical Science*, 246, 645–56.
Hurd, J. P. (1985) 'Kissing cousins. Frequencies of cousin types in Nebraska Amish marriages', *Social Biology*, 32 (1–2), 82–9.
Ingold, T. (1992) 'Culture and perception of the environment', in E. Croll and D. Parkin (eds) *Bush Base: Forest Farm Culture, Environment and Development*, Routledge, London.
Inoguchi, R. and Nakajima, T. (1960) *The Divine Wind: Japan's Kamikazi Force in World War Two*, Bantan, New York.
Isaac, E. (1961–62) 'The act and the covenant; the impact of religion on the landscape', *Landscape*, 11(2), 12–17.
Isbell, B. J. (1978) *To Defend Ourselves: Ecology and Ritual in an Andean Village*, Institute of Latin American Studies, University of Texas Press, Austin.
Jacobi, H. (1914) 'Jainism', in W. Hastings, (ed.) *Encyclopedia of Religion and Ethics*, Clark, Edinburgh.
Joseph, M. (1917) 'Care of the aged', in W. Hastings, (ed.) *Encyclopedia of Ethics and Religion*, Clark, Edinburgh.
Juergensmayer, M. (1993) *The New Cold War: Religious Nationalism Confronts the Secular State*, University of California Press, Berkeley.
Kakar, S. (1990) 'Some unconscious aspects of ethnic violence in India', in V. Das, (ed.) *Communities, Riots and Survivors in South Asia*, Oxford University Press, Delhi, 135–45.
Kaku, K. (1975a) 'Increased induced abortion rate in 1966: an aspect of Japanese folk superstition', *Annals Human Biology*, 2, 111–15.
Kaku, K. (1975b) 'Were girl babies sacrificed to a folk superstition in 1966 in Japan?', *Annals Human Biology*, 2, 391–3.
Kazemzadeh, H. K. Z. (1912) *Relation d'un Pèlerinage à La Mecque en 1910–11*, Leroux, Paris.
Kenyon, S. M. (1995) 'Zar as modernisation in central Sudan', *Anthropological Quarterly*, 68(2), 107–20.
Keuneman, H. (1992) *Sri Lankha*, Apa Publications, Colombo.
Keys, A. (1961) 'Further observations on monastic cholesterol', *New England Journal of Medicine*, 264, 1005.

Kimambo, I. N. and Omari, C. K. (1972) 'The development of religious thought and centres among the Pare', in T. O. Ranger, and I. M. Kimambo, (eds) *The Historical Study of African Religion*, Heinemann, London, 111–21.
Kindere, L. V. (1904) *La Chronique de Gislebert de Mons*, Brussels.
Kleinman, A. (1980) *Patients and Healers in the Context of Culture*, California University Press, Berkeley.
Klijn, A. F. L. (1962) *The Apocryphal Acts of Thomas*, University of Leiden Press, Leiden.
Knight, H. (1954) *Food Administration in India 1939–47*, Stanford University Press, Stanford, Calif.
Knutsson, K. E. and Selinus, R. (1970) 'Fasting in Ethiopia: an anthropological and nutritional study', *American Journal of Clinical Nutrition*, 23, 956–9.
Kochar, V. (1977a) 'Sanitation and culture, 1. Social aspects of sanitation and personal hygiene in a rural West Bengal region', *Indian Journal of Preventative and Social Medicine*, 8, 106–17.
Kochar, V. (1977b) 'Sanitation and culture, 2. Behavioural aspects of the disposal of excreta in a rural West Bengal region', *Indian Journal of Preventative and Social Medicine*, 8, 142–51.
Kochar, V. (1978) 'Human factors in hookworm transmission in a rural West Bengal region', *Indian Journal of Preventative and Social Medicine*, 9 (3).
Kochar, V. K. et al. (1976) 'Human factors in the regulation of parasitic infections: cultural ecology of hookworm population in rural West Bengal', in F. X. Grollig, and H. B. Haley, (eds) *Medical Anthropology*, Mouton, The Hague, 287–312.
Krantz, G. S. (1970) 'Human activities and megafaunal extinction', *American Scientist*, 58, 164–70.
Krech, S. (1999) *The Ecological Indian Myth and History*, Norton, New York.
Lachiver, M. and Dupaquier, J. (1969) 'Breast feeding and fertility', *Annales: Économie, Societé, Civilisation*, 1399.
Laidlaw, J. (1995) *Riches and Renunciation: Religion, Economics and Society among the Jains*, Oxford University Press,Oxford.
Lamb, T. (1932) *Tamurlane, the Earth Shaker*, New York.
Lambo, T. A. (1971) *Problems of Adjustment between Traditional and Modern Methods of Medical Practice*, University of Ibadan.
Lane, B. C. (1988) *Landscapes of the Sacred: Geography and Narrative in American Spirituality*. Paulist Press, New York.
Lansing, J. S. (1987) 'Balinese water temples and the management of irrigation', *American Anthropologist*, 89(2), 326–41.
Lansing, J. S. (1991) 'Priests and Programmes: Technologies of Power in the Engineered Landscape of Bali. Princeton University Press, Princeton, New Jersey.
Larson, D. et al. (1989) 'The impact of religion on men's blood pressure', *Journal of Religion and Health*, 28, 265–78.
Lasaga, J. I. (1980) 'Death in Jonestown: techniques of political control of a paranoid leader', *Suicide Life-Threatening Behaviour*, 10, 210–13.
Lawrence, R. J. (1996) 'The multi-dimensional nature of boundaries: an integrative perspective', in D. Fellow, (ed.) *Setting Boundaries: the Anthropology of Spatial and Social Organisation*, Bergin and Garvey, Westport.
Leach, E. (1959) 'Hydraulic society in Ceylon', *Past and Present*, 15.
Leadaam, I. S. (1897) *The Doomsday of Inclosures: 1517–18*. Longmans, London.

Lemon, F. R. et al. (1964) 'Cancer of the lung and mouth in Seventh Day Adventists', *Cancer*, 17, 486–97.
Levin, J. S. (1996) 'How religion influences morbidity and health: reflections on natural history, salutogenesis and host resistance', *Social Science and Medicine*, 43 (5), 849–64.
Levin, J. S. and Vanderpool, N. Y. (1989) 'Is religion therapeutically significant for hypertension', *Social Science and Medicine*, 29 (1), 69–78.
Levine, R. R. (1968) 'Judaism and some modern medical problems', *Journal of the Medical Society of New Jersey*, 65, 638–9.
Levy, G. (1974) *Religion and Revolution*, Oxford University Press, New York.
Livingstone, D. (1857) *Missionary Travels in South Africa*, John Murray, London.
Lodrick, D. O. (1981) *Sacred Cows, Sacred Places: Origins and Survivals of Animal Homes in India*, University of California Press, Berkeley.
Loftin, J. D. (1991) *Religion and Hopi Life in the 20th Century*, Indiana University Press, Bloomington.
Loizos, P. (1988) 'Intercommunal killing in Cyprus', *Man* (n.s.), 23, 634–53.
Lokhandwala, S. T. (1985) 'Indian Islam: composite culture and integration', *New Quest*, March–April, 87–101.
Losito, P. (1946) 'Church hygiene: microbic contents of holy water in various churches in Perugia', *Bull. Soc. Ital. Bio. Sp*, 22, 463–5.
Ludwig, E. (1936) *The Nile: the Life Story of a River*, M. H. Lindsay, (trans.), Allen and Unwin, London.
Macinnis, D. E. (1973) *Religious Policy and Practice in Communist China*, Macmillan – now Palgrave, New York.
Mackenbach, J. P. et al. (1993) 'Self-reported morbidity and disability among Trappist and Benedictine monks', *American Journal of Epidemiology*, 138 (8), 569–73.
McEvoy, L. et al. (1981) 'Life style and death patterns of the Missouri Latter Day Saints members', *American Journal of Public Health*, 71 (12), 1350–7.
McGovern, T. H. (1980) 'Adaptation and extinction in Norse Greenland. Cows, harp seals and church bells', *Human Ecology*, 8(3), 245–75.
McGuire, R. B. (1985) *Ritual Healing in Suburban America*, Rutgers University Press, New Brunswick.
McKusick, V. A. (1973) 'Genetic studies in American inbred populations with particular reference to the Old Order Amish', *Israeli Journal of Medical Science*, 9, 1276–84.
McMillen, S. I. (1971) *None of These Diseases*, Lakeland, California.
McNaspy, C. J. (1982) *The Lost Cities of Paraguay: Art and Architecture of the Jesuit Reductions*, Loyola University Press, Chicago.
McNeil, J. T. and Gamer, H. M. (eds) (1938) *Medieval Handbooks of Penance*, Columbia University Press, New York.
Maddison, D. and Viola, A. (1968) 'The health of widows in the year following bereavement', *Journal of Psychosomatic Research*, 12, 297–306.
Mandelbaum, D. G. (1941) 'Cultural changes among the Nilgiri tribes', *American Anthropologist*, 43, 19–26.
Manley, G. T. and Neech, A. S. (1950) 'Hinduism', in J. N. D. Anderson (ed.) *The World's Religions*, Inter-Varsity Fellowship, London, 99–117.
Manu, Laws of (1967) G. Buhkler (trans.), Dover, New York.
Marnham, P. (1980) *Lourdes: a Modern Pilgrimage*, Heinemann, London.

Marsh, G. P. (1847) *Address Delivered before the Agricultural Society of Rutland Court, Vermont*, 30 September.

Marshall, P. (1994) *The Catholic Priesthood and the English Reformation*, Clarendon Press, Oxford.

Maslow, A. H. (1987) *Motivation and Personality*, Harper and Row, New York.

Maspero, H. (1950) *Les religions chinoises: Le Taoisme, Études Historique*, Civilisations du Sud, Paris.

Mayadas, C. (1954) *Between Us and Hunger*, Oxford University Press, London.

Meller, H. (1999) *London Cemeteries*, Ashgate, Aldershot.

Menendez, A. J. (1973) *The Bitter Harvest: Church and State in Northern Ireland*, Luce, Washington.

Mercier, C. A. (1915) *Leper Houses and Medieval Hospitals*, Lewis, London.

Merton, R. K. (1964) *Social Theory and Social Structure*, Free Press, London.

Mesaki, S. (1992) 'The preponderance of women as victims in Sukuma witch killings,1965–6', in P. G. Foster and S. Maghimbi (eds) *The Tanzanian Peasantry; Economy in Crisis*, Avebury, Aldershot.

Meslin, M. (1972) 'Le phénomène religieux populaire', in B. Lacroix and P. Boglioni (eds) *Les Religions Populaires*, Colloque International, Quebec.

Michael, F. (1966) *The Taiping Rebellion*, University of Washington Press, Seattle.

Midgely, M. (1992) 'Towards a more humane view of the beasts'? in D. E. Cooper and J. A. Palmer (eds) *The Environment in Question*, Routledge, London, 28–36.

Milner, M. (1994) 'Status and sacredness: worship and salvation as forms of status transformation', *Journal of the Scientific Study of Religion*, 33 (2), 99–109.

Mirone, L. (1954) 'Nutrient intake and blood findings in a diet devoid of meat', *American Journal of Clinical Nutrition*, 2, 246.

Mishkat-al-Masabih (1963) *A Collection of the Most Authentic Traditions Regarding the Actions and Sayings of Muhammad*, J. Robson (trans.), Ashraf, Lahore.

Mitchell, C. W. (1993) 'Islamic beliefs and agricultural development programmes', in J. Hole and B. Schantz (eds) *The Three Angels and the Crescent*, Newbold College, Bracknell, 83–93.

Mitford, N. (1963) *The American Way of Life*, Simon and Schuster, New York.

Morenz, S. (1994) *Egyptian Religion*, A. E. Keep (trans.), Cornell University Press.

Moule, A. C. (1957) *Quinsai with Other Notes on Marco Polo*, Cambridge University Press, Cambridge.

Mubarak, A. (1887–89) *Memoirs*, 20 volumes, Cairo.

Mukerjee, R. (1938) *The Regional Balance of Man: an Ecological Theory of Population*, University of Madras Press, Madras.

Murra, J. V. (1965) 'Herds and herdsmen of the Inca state', in A. Leeds and A. P. Vayda (eds) *Man, Culture and Animals*, American Association for the Advancement of Science, Washington DC.

Mus P. (1936) 'Le symbolisme à Angkor Thom: le grand miracle du Bayon', Académie des Inscriptions et Belles Lettres, *Comptes Rendus du Séance*, 57–68.

Nandia, P. N. et al. (1955) *Report of the Expert Committee on the Prevention of Slaughter of Cattle in India*, Government of India Press, New Delhi.

Nandy, A. (1990) 'The politics of secularism and the recovery of religious tolerance', in V. Das (ed.) *Mirrors of Violence*, Oxford University Press, Delhi.

Nansen, F. (1893) *Eskimo Life*, Longman, London.

Nash, J. (1975) *We Eat the Mines and the Mines Eat Us: Dependency and Exploitation in Bolivian Tin Mines*, Columbia University Press, New York.

Nayanar, C. (1953) *The Religion of Ahimsa*, New Delhi.

Needham, J. (1954) *Science and Civilisation in China*, Vol. 1, Cambridge University Press, Cambridge.

Needham, R. (1972) *Belief, Language and Experience*, Blackwell, Oxford.

Nelson, H. G. H. (1974) 'Ancestor worship and burial practices', in A. P. Wolf (ed.) *Religion and Ritual in Chinese Society*, Stanford University Press, Stanford, Calif., 251–77.

Newsome, A. E. (1980) 'The economic theology of the Red Kangaroo in Central Australia', *Mankind*, 12, 327–33.

Nodari, R. and Pirovane G. (1970) 'Sulla Mortalite: Raffronte Dei Dati Statistici Referiti Ai Bambini Nati Nel Matrimonie E Buoxi Del Matrimonie', *Minerva Pediatrica*, 22, 1254–6, 1449–51.

Nolan, M. L. (1991) *The European Roots of Latin-American Pilgrimage*, Greenwood Press, New York.

Nolan, M. L. and Nolan, S. (1989) *Christian Pilgrimage in Modern Western Europe*, University of North Carolina Press, Chapel Hill.

Noonan, J. T. (1966) *Contraception: a History of its Treatment by Catholic Theologians and Canonists*, Harvard University Press, Cambridge, Mass.

Noss, J. B. (1959) *Man's religion*, Macmillan – now Palgrave, New York.

Novotny, T. et al. (1988) 'Measles outbreaks in religious groups exempt for immunisation laws', *Public Health Review*, 103 (1), 49–54.

Odend'hal, S. (1972) 'Cross energetic efficiency of Indian cattle in their environment', *Journal of Human Ecology*, 1–27.

O'Farrell, P. (1971) *Ireland's English Question*, Schocken, New York.

O'Hara, M. L. (1980) *An Unfinished Canticle: the Story of the Franciscan Missionary Sisters of Africa*, St Joseph's College Archives, Mill Hill.

Ohsawa, G. (1965) *Zen Macrobiotics: the Art of Longevity and Rejuvenation*, Ohsawa Foundation, New York.

Oliver, S. (1894) *Historia Damiatana: Bibliothek Dei Literarischen Verens in Stuttgart*, Tübingen.

Olivier, L. and Ansari, N. (1967) 'Epidemiology of bilharziasis', in F. K. Mostofik (ed.) *Bilharziasis*, Springer Verlag, New York.

O'Malley, P. (1983) *The Uncivil War: Ireland Today*, Houghton Mifflin, Boston.

Orni, E. and Efrat, E. (1964) *The Geography of Israel*, Israel Program for Scientific Translations, Jerusalem, distributed by Oldbourne Press, London.

Orsi, R. (1985) *The Madonna of 115th Street: Faith and Community in Italian Harlem,1880–1950*, Yale University Press, Cambridge, Mass.

Ortiz, A. (1969) *The Tewa World: Space, Time, Being and Becoming in a Pueblo Society*, University of Chicago Press, Chicago.

Orubuloye, I. O. et al. (1994) 'The role of religious leaders in changing sexual behaviour in South-West Nigeria in an era of AIDS', in I. O. Orubuloye et al. (eds) *Sexual Networks and Aids in Sub-saharan Africa*, Australian National University Press, Canberra.

Pakrasi, K. B. (1970) *Female Infanticide in India*, Editions Indian, Calcutta.

Palmer, M. (1998) 'Chinese religion and ecology', in D. E. Cooper and J. A. Palmer (eds) *Spirit of the Environment: Religion, Value and Environmental Concern*, Routledge, London, 15–29.

Pandey, G. (1990) 'The colonial construction of "communalism": British writings on Banaras in the 19th century', in V. Das (ed.) *Communities, Riots and Survivors in South Asia*, Oxford University Press, Delhi.

Pannikar, R. (1979) *The Vedic Experience*, Darton, Longman & Todd, London.

Panter-Brick, C. (1991) 'Parental responses to consanguinity and genetic disease in Saudi Arabia', *Social Science and Medicine*, 33 (11),1295–302.

Parish, W. L. and Whyte, M. K. (1978) *Village and Family in Contemporary China*, University Chicago Press, Chicago.

Parker, J. (1961) 'Out of wedlock births in New York City: II. Medical aspects', *American Journal of Public Health*, 51, 846–55.

Parrack, D. W. (1969) 'An appraisal to the bioenergetics of rural West Bengal', in P. A. Vayda (ed.) *Environment and Cultural Behaviour*, Natural History Press, Garden City, New Jersey.

Parry, J. P. (1994) *Death in Banaras*, Cambridge University Press, Cambridge.

Patimokkha, The (1966) *The Monastic Rules in Buddhism*, Social Science Association Press of Thailand, Bangkok.

Peel, J. D. Y. (1968) *Aladura: a Religious Movement among the Yoruba*, Oxford University Press, London.

Peires, J. B. (1989) *The Dead Will Arise: Nongqawuse and the Great Cattle Killing Movement of 1856–7*, Raven Press, Johannesburg.

Pemberton, J. (1994) *On the Subject of Java*, Cornell University Press, Ithaca.

Phillips, R. L. (1975) 'Role of life-style and dietary habits in risk of cancer among Seventh Day Adventists', *Cancer Research*, 35, 3513–22.

Phillips, R. L. (1980) 'Cancer among Seventh day Adventists', *Journal of Environmental Pathology and Toxicology*, 3, 157–69.

Philp, H. R. A. (1936) *A New Day in Kenya*, London.

Piggott, J. (1969) *Japanese Mythology*, Hamlyn, London.

Pittin, R. (1996) 'Negotiating boundaries: a perspective from Nigeria', in D. Pellow (ed.) *Setting Boundaries: the Anthropology of Spatial and Social Organisation*, Borgen and Garvey, Westport.

Pollitzer, R. and Swaroop, S. (1959) *Cholera*, World Health Organisation Monograph Series 43, Geneva.

Postan, M. M. (1966) *Agrarian Life in the Middle Ages*, Cambridge Economic History of Europe, Cambridge University Press, Cambridge.

Potter, R. G. et al. (1964) *Application of Field Studies to Research on the Physiology of Human Reproduction: Lactation and its Effects upon Birth Interval in Eleven Punjab Villages: Symposium on Research Issues in Public Health and Population Change*, University of Pittsburgh, Pittsburgh.

Pousin, L. de la V. (1917) *The Way of Nirvana*, Cambridge University Press, Cambridge.

Pritchard, J. (ed.) (1955) *Ancient Near-eastern Texts*, Princeton University Press.

Puzo, M. (1972) *The Godfather*, Book Club, London.

Quint, M. N. (1969) 'The idea of progress in an Iraqi village', in J. M. Landau (ed.) *Man, State and Society in the Contemporary Middle East*, Pall Mall Press, London.

Rackham, O. (1986) *The History of the Countryside*, Dent, London.

Rafeq, A.-K. (1966) *The Province of Damascus, 1723–83*, Khayats, Beirut.

Ranger, T. O. (1981) 'Godly medicine: the ambiguities of medical missions in South-East Tanzania,1900–45', *Social Science and Medicine*, 15 (8), 261–77.

Rao, P. S. (1983) 'Religion and intensity of inbreeding in Tamil Nadu, South India', *Social Biology*, 309 (4), 413–22.
Rapoport, D. C. (1982) 'Terror and the Messiah: an ancient experience and some modern parallels', in D. C. Rapoport and T. Alexander (eds) *The Morality of Terrorism: Religion and Secular Justification*, Pergamon Press, New York.
Rappaport, R. A. (1967) *Pigs for Ancestors: Ritual in the Ecology of a New Guinea People*, Yale University Press, New Haven.
Rasmussen, K. (1931) *The Netsilik Eskimo, Report of the Fifth Thule Expedition, 1921–4*, Gylendalske Boghandel, Copenhagen.
Ray, D. J. (1992) *The Eskimos of the Bering Strait, 1650–1898*, University of Washington Press, Seattle.
Reddy, D. V. S. (1941) 'Medical relief in medieval South India: centres of medical aid and types of medical institutions', *Bulletin History of Medicine*, 9, 385–400.
Rees, W. D. and Lutkins, S. G. (1967) 'The mortality of bereavement', *British Medical Journal*, 4, 13–16.
Rewell, R. E. (1957) 'Ethnological factors in the aetiology of cancer of the uterine cervix', *Journal of Obstetrics and Gynaecology, British Empire*, 64, 821–6.
Reynolds, V. and Tanner, R. E. S. (1983) *The Biology of Religion*, Longman, London.
Reynolds, V. and Tanner, R. E. S. (1995) *The Social Ecology of Religion*, Oxford University Press, New York.
Ricard, R. (1966) *The Spiritual Conquest of Mexico*, University California Press, Berkeley.
Richards, I. A. (1953) 'Towards a theory of translating', in A. F. Wright (ed.) *Studies in Chinese Thought*, University of Chicago Press, Chicago, 247–62.
Richardson, I. F. (1950) 'Statistics of deadly quarrels', in T. H. Pear (ed.) *Psychological Factors of Peace and War*, Hutchinson, London, 237–55.
Risso, P. (1995) *Merchants and Faith: Muslim Commerce and Culture in the Indian Ocean*, Westview Press, Boulder.
Robbins, J. (1991) *Diet for a New America*, Video. Earthsave Foundation, Felton, California.
Roberts, D. F. and Tanner, R. E. S. (1959) 'A demographic study in an area of low fertility in North-East Tanganyika', *Population*, 13 (1), July, 321–46.
Roberts, I. F. et al. (1979) 'Malnutrition in infants receiving cult diets: a form of child abuse', *British Medical Journal*, 1, 296–8.
Robinson, R. H. (1959) 'Buddhism in China and Japan', in R. C. Zaehner (ed.) *The Encyclopedia of Living Faiths*, Hutchinson, London, 321–46.
Rogers, L. (1926) 'The conditions influencing the incidence and spread of cholera in India', *Proceedings Royal Society Medicine, Epidemic Section*, 19, 59–91.
Rosen, G. (1958) *A History of Public Health*, MD Publications, New York.
Rosny, L. de (1837) *Histoire de l'Abbaye de Notre Dame de Loos*, Lille and Paris.
Ross, J. L. (1984) 'Culture and fertility in the Nepal Himalayas', *Human Ecology*, 12 (2), 163–81.
Roth, D. M. (1977) *The Friar Estates in the Philippines*, University of New Mexico Press, Albuquerque.
Roy, P. (1955) 'The sacred cow in India', *Rural Sociology*, 20, 8–15.
Roy, T. K. et al. (1991). 'Education, fertility and contraception among Hindus and Roman Catholics in Goa', *J. Biosocial Science*, 23(3), 353–8.
Rudner, D. W. (1994) *Caste and Capitalism in Colonial India: the Nattukottai Chettiars*, University of California Press, Berkeley.

Rupesinghe, K. (1994) 'Forms of violence and its transformation', in K. Rupesinghe and M. C. Rubio (eds) *The Culture of Violence*, United Nations University Press, Tokyo.

Russell, C. and Russell, W. M. S. (1968) *Violence: Monkeys and Man*, Macmillan – now Palgrave, London.

Sachar, H. M. (1987) *A History of Israel: 2. From the Aftermath of the Yom Kippur War*, Oxford University Press, Oxford.

Sallnow, M. (1987) *Pilgrims of the Andes Regional Cults in Cuzco*, Smithsonian Institute Press, Washington, DC.

Sangren, P. S. (1987) *History and Magic Power in a Chinese Community*, Stanford University Press.

Sanmartin, R. (1996) 'The environment and the cultural system', in J. van Bremen et al. (eds) *Horizons of Understanding*, University of Leiden Press, Leiden, 244–57.

Sapir, E. (1949) 'Cultural anthropology and psychiatry', in D. Mandelbaum (ed.) *Selected writings in Language, Culture and Personality*, University of California Press, Berkeley, 160–8.

Saraswati, B. (1975) *Kasli: Myth and Reality of a Classical Cultural Tradition*, Indian Institute of Advanced Studies, Simla.

Sattin, A. (ed.) (1986) *An Englishwoman in India: the Memoirs of Harriet Tytler. 1828–58*, Oxford University Press, Oxford.

Schafer, E. H. (1962) 'The conservation of nature under the T'ang dynasty', *Journal of Economic and Social History of the Orient*, 5, 280–1.

Schapera, I. (1966) *Married Life in an African Tribe*, Faber, London.

Schwarz, B. E. (1960) 'Ordeal by serpents, fire and strychnine: a study of some provocative psychosomatic phenomena', *Psychiatry Quarterly*, 34, 405–29.

Schweitzer, A. (1963) *The Primeval Forest*, Pyramid, New York.

Schweitzer, A. (1965) *The First Hospital at Lambaréné*, Holt, New York.

Scurletis, T. D. et al. (1969) 'Trends in illegitimacy and associated mortality in North Carolina, 1957–66', *North Carolina Medical Journal*, 10, 214–21.

Seguy, J. (1973) 'Religion and agricultural success: the vocational life of the French Anabaptists from 17th to 19th centuries', M. Shank (trans.), *Mennonite Quarterly*, R. 47.

Semple, E. C. (1947) *Influences of Geographic Environment on the Basis of Ratzel's System of Anthropo-geography*, Henry Holt, New York.

Sen, H. (1903) 'Cholera in Puri with a special account of the epidemic during 1901', *Indian Medical Gazette*, 38, 135–8.

Shaheed, F. (1990) 'The Pathan–Muhajin conflicts,1985–6: a national perspective', in V. Das (ed.) *Communities, Riots and Survivors in South Asia*, Oxford University Press, Delhi, 194–214.

Sharot, S. (1996) 'Traditional, modern or post-modern? Recent religious developments among Jews in Israel', in K. Flanagan and P. C. Jupp (eds) *Postmodernity, Sociology and Religion*, Macmillan – now Palgrave, Basingstoke.

Sharp, L. (1952) 'Steel axes for stone age Australians', *Human Organisation*, 11 (2), 457–64.

Shastri, C. P. (1960) 'Bullock labour utilization in agriculture', *Economic Weekly*, 12, 1585–92.

Sheffield, C. (1983) *Man on Earth*, Sidgwick and Jackson, London.

Sherring, M. A. (1872) *Hindu Tribes and Castes*, Thacker Spink, Calcutta.

Sima, Quan (135–45 bc) *Records of the Historian: Biography of the Emperor Qui-shi-Huang*. Quoted in Wu-Xiaolong. (1999). *The subterranean Army of Emperor Qui-Shi-Huang*. China International Tourism,Beijing.
Simoons, F. J. (1961) *Eat Not This Flesh: Food Avoidances in the Old World*, University of Wisconsin Press, Madison.
Simoons, F. J. (1968) *A Ceremonial Ox of India: the Mithan in Nature, Culture and History*, University of Wisconsin Press, Madison.
Sims, G. R. (ed.) (1906) *Living London*, Cassell, London..
Singh, R. C. (1955) *Banaras: a Study in Urban Geography*, Nand Kishor, Banaras.
Skar, S. L. (1994) *Lives Together – Worlds Apart: Quechua Colonization in Jungle and City*, Scandinavian University Press, Stockholm.
Slack, P. (1985) *The Impact of the Plague in Tudor and Stuart England*, Routledge, London.
Smith, H. S. (1969) 'Animal domestication and animal cult in Dynastic Egypt', in P. J. Ucko and G. W. Dimbleby (eds) *The Domestication and Exploitation of Plants and Animals*, Duckworth, London, 313.
Smith, W. R. (1901) *Lectures on the Religion of the Semites*, Black, London.
Smith, W. R. (1977) *The Fiesta System and Economic Change*, Columbia University Press, New York.
Social Science and Immunisation Research Project (1988). Centre for Development Economics, University of Delhi, Delhi.
Soga, T. (1857) *Comment on the Xhosa Cattle Killing Movement*, UPC Mission Record.
Solomon, N. (1994) 'Judaism', in J. Holm (ed.) *Attitudes to Nature*, Pinter, London.
Southwold, M. (1982) 'True Buddhism and village Buddhism in Sri Lankha', in J. Davis (ed.) *Religious Organisation and Religious Experience*, Academic Press, London, 140–9.
Spier, F. (1992) *Religious Regimes in Peru: Religion and State Development in Long-Term Perspective and the Effects in the Andean Village of Zurite*, Amsterdam University Press, Amsterdam.
Spiro, M. E. (1967) *Burmese Supernaturalism: a Study in the Explanation and Reduction of Suffering*, Prentice-Hall, New Jersey.
Spiro, M. E. (1971) *Buddhism and Society: A Great Tradition and its Burmese Vicissitudes*, Allen & Unwin, London.
Spiro, M. E. (1977) *Kinship and Marriage in Burma*, University of California Press, Berkeley.
Sprigg, J. (1975) *By Shaker Hands*, Knopf, New York.
Srinivas, M. N. (1958) 'India's cultural values and economic development', *Economic Development and Cultural Change*, 7, 3–6.
Srinivas, M. N. (1987) *The Dominant Caste and Other Essays*, Oxford University Press, Delhi.
Steiner, G. (1975) *After Babel: Aspects of Language and Translation*, Oxford University Press. Oxford.
Stevenson, H. N. C. (1943) *The Economics of the Central Chin Tribes*, Times of India Press, Bombay.
Steward, P. J. (1978) 'Islamic law as a factor in grazing management. The pilgrimage sacrifice', *Proceedings First International Rangeland Congress*, Denver, Colorado.

Stover, L. E. (1974) *The Cultural Ecology of Chinese Civilisation: Peasants and Elites in the Last of the Agrarian States*, Pica, New York.

Strikman, M. (1982) 'India in the Chinese looking glass', in Klimburg-Salter (ed.) *The Silk Route and the Dharma Path: Esoteric Buddhist Art on the Trans-Himalayan Trade Routes*, Los Angeles.

Surbey, M. K. (1986) 'Seasonality of conception in Hutterite colonies of Europe (1758–1881) and North America (1858–1964)', *Journal Biosocial Science*, 18 (3), 337–45

Swaroop, S. and Raman, M. V.(1951) 'Endemics of cholera in relation to fairs and festivals in India', *Indian J. Medical Research*, 39, 41–9.

Takeda, C. (1976) 'Family religion in Japan: the "ie" and its religious faith', in A. Bharati (ed.) *The Realm of the Extra-human: Ideas and Action*, Mouton, Le Hague.

Tanner, A. (1979) *Bringing Home Animals*, Hurst, London.

Tanner, R. E. S. (1957) 'The installation of Sukuma chiefs in Mwanza district, Tanganyika', *African Studies*, 16 (4), 197–209.

Tanner, R. E. S. (1958a) 'Ancestor propitiation ceremonies in Sukumaland', *Africa*, 28, 225–36.

Tanner, R. E. S. (1958b) 'Sukuma ancestor worship and its relationship to social structure', *Tanganyika Notes and Records*, 50, June, 52–62.

Tanner, R. E. S. (1964) 'Cousin marriage in the Afro-Arab community of Mombasa, Kenya', *Africa*, 34, 127–38.

Tanner, R. E. S. (1969) *Transition in African Belief*, Maryknoll, New York.

Tanner, R. E. S. (1975) 'A note on Christianity and traditional religious practices in Ngara, Tanzania', *Journal of Religion in Africa*, 6, 161–6.

Tanner, R. E. S. (1995) 'Excreting, excretors and social health policy: some cross-cultural observations on under-research activity', *Journal of Preventive Medicine and Hygiene*, 36 (3/4), 85–94.

Tanner, R. E. S. (2000) 'Violence and the natural environment', *Journal of Social Sciences*, 4 (1), 101–9.

Tapia, A. de (1866) 'Relacion sobre la Conquista de Mexico', in J. G. Icazbalceta, *Colleccion De Documentos Pasa La Historia De Mexico*, Mexico City.

Thomas, E. S. (1920) 'State of the dead: Buddhist', in J. Hastings (ed.) *Encyclopedia of Religion and Ethics*, Clark, Edinburgh, II, 829–33.

Toelken, B. (1976) 'How many sheep will it hold', in W. H. Capps (ed.) *Seeing with a Native Eye: Essays on Native American Religion*, Harper and Row, New York, 9–24.

Toland, J. (1971) *The Rising Sun: the Decline and Fall of the Japanese Empire, 1936–45*, Cassell, London.

Topley, M. and Hayes, J. (1968) 'Notes on some Vegetarian Halls in Hong Kong belonging to the sect of Hsien-T'ien (the way of former heaven)', *Journal of Hong Kong Branch Royal Asiatic Socety*, 8, 135–48.

Troyer, H. (1988) 'Review of cancer among four religious groups: evidence that life-styles are distinctive sets of risk factors', *Social Science and Medicine*, 26 (10), 1007–17.

Tuan, Y.-F. (1974) *Topophilia: a Study of Environmental Perception, Attitude and Values*, Columbia University Press, New York.

Turnbull, C. (1961) *The Forest People: a Study of the Pygmies of the Congo*, Simon and Schuster, New York.

Turnbull, C. (1966) *Wayward Servants: The World of the African Pygmy*, Eyre and Spottiswoode, London.

Turnbull, C. (1978) 'Rethinking the Ik: a functional non-social system', in C. D. Laughlin and I. A. Brady (eds) *Extinction and Survival in Human Populations*, Columbia University Press, New York. 49–75.

Turnbull, C. (1994) *The Mountain People*, Pimlico, London.

Turner, H. W. (1967) *African Independent Church: 2.* Oxford University Press, London.

Turner, V. (1974) *Dramas, Fields and Metaphors: Symbolic Action in Human Society*, Cornell University Press, Ithaca.

Turner, V. and Turner, E. (1978) *Image and Pilgrimage in Christian Culture*, Blackwell, Oxford.

Ukaegbu, A. O. (1977) 'Fertility of women in polygynous unions in rural Eastern Nigeria', *Journal of Marriage and the Family*, 39, 397–404.

Underwood, P. and Underwood, Z. (1980) New spells for old: expectations and realities of Western medicine in a remote tribal society in Yemen, Arabia, in N. F. Stanley and J. R. Joshe (eds) *Changing Disease Patterns and Human Behaviour*, Academic Press, New York, 294.

Vanderpool, H. Y. (1986) The Wesleyan-Methodist tradition, in R. L. Numbers and D. W. Amundsen (eds) *Caring and Curing: Health and Medicine in the Western Religious Tradition*, Macmillan – now Palgrave, New York.

Vaux, K. (1976) 'Religion and health', *Preventative Medicine*, 5 (4), 522–36.

Veer, Pn der (1994) *Religious Nationalism: Hindus and Muslims in India*, University of California Press, Berkeley.

Vogt, E. and Albert, E. M. (eds) (1966) *The People of Rimrock: a Study of Values in Five Different Cultures*, Harvard University Press, Cambridge, Mass.

Voland, E. (1988) 'Differential infant and child mortality in evolutionary perspective: data from late 17th to 19th century Ostfriesland (Germany)', in L. L. Betzig et al. (eds) *Human Reproduction: a Darwinian Perspective*, Cambridge University Press, Cambridge, 392–400.

Waddell, L. A. (1967) *The Buddhism of Tibet*, Heffer, Cambridge.

Waghorne, J. P. (1999) 'The diaspora of the Gods: Hindu temples in the New World system 1640–1800', *Journal of Asian Studies*, 58 (3), 648–86.

Wakeman, F. (1966) *Strangers at the Gate: Social Disorder in South China,1839–61*, University of California Press, Berkeley.

Wallerstein, I. (1991) *Geopolitics and Geoculture: Essays on the Changing World System*, Cambridge University Press, Cambridge.

Walsh, J. J. (1978) *The Catholic Church and Healing*, Burns, Oates and Washbourne, London.

Watt, W. M. (1953) *Muhammad at Mecca*, Clarendon Press, Oxford.

Watt, W. M. (1956) *Muhammad at Medina*, Clarendon Press. Oxford.

Waugh, E. (1956) *The Loved One*, Chapman and Hall, London.

Wei-Ming, T. (1994) 'Beyond the Enlightenment mentality', in M. E. Tucker and J. A. Grim (eds) *World Views and Ecology: Religion, Philosophy and the Environment*, Orbis, Maryknoll, New York.

Weiner, I. et al. (1951) 'Carcinoma of the cervix in Jewish women', *American Journal of Obstetrics and Gynaecology*, 61, 418–22.

Weiss, N. (1888) 'La Seine et le nombre des victimes parisiennes de la Saint Barthélemy', *Bulletin de la Societé de l'histoire du Protestantisme Français*, 36, 341–74.

Welbourn, F. and Ogot, B. A. (1966) *A Place to Feel at Home: a Study of Two Independent Churches in Kenya*, Oxford University Press, London.

Welch, H. (1967) *The Practice of Chinese Buddhism, 1900–50,* Harvard University Press, Cambridge, Mass.

Werbner, R. P. (1989) *Ritual Passage, Sacred Journey: the Process and Organisation of Religious Movement*, Smithsonian Institute, Washington DC.

Werner, D. (1983) 'Fertility and pacification among the Mekranoti of Central Brazil', *Human Ecology,* 11 (2), 227–45.

West, D. W. et al. (1980) 'Cancer risk factors: an analysis of Utah Mormons and non-Mormons', *Journal of the National Cancer Institute,* 65, 1083–95.

Weyl, N. (1968) 'Some possible genetic implications of Carthaginian child sacrifice', *Perspectives in Biological Medicine,* 12, 69–78.

Wheatley, P. (1971) *The Pivot of the Four Quarters: a Preliminary Enquiry into the Origins and Character of the Ancient Chinese City,* Edinburgh University Press, Edinburgh.

Whorf, B. L. (1956) *Language, Thought and Reality,* Princeton, Cambridge, Mass

Wilkinson, D. (1980) *Deadly Quarrels*, University of California Press, Berkeley.

Willcocks, W. (1935) *Sixty Years in the East,* Blackwood, Edinburgh.

Wilson, G. E. (1956) 'Christian Science and longevity', *Journal of Forensic Science,* 1, 43–60.

Winkelman, M. (1998) 'Aztec human sacrifice: cross-cultural assessments of the ecological hypothesis', *Ethnology,* 37 (3), 285–98..

Wolf, E. R. (1971) 'Peasant rebellion and revolution', in N. Miller and R. Aya (eds) *National Liberation,* Free Press, New York, 50–61.

Wood, I. F. (1920) 'The state of the dead' in J. Hastings (ed.) *Encyclopedia of Religion and Ethics,* Clark, Edinburgh, 841–3.

Wright, Q. (1964) *A Study of War,* University of Chicago Press, Chicago.

Wuthnow, R. A. (1977) 'A longitudinal, cross-national indicator of societal–religious commitment', *Journal of the Scientific Study of Religion,* 16, 87–99.

Wynder, E. L. et al.(1954) 'A study of environmental factors in carcinoma of the cervix', *American Journal of Obstetrics and Gynaecology,* 68, 1016–52.

Wynder, E. L. et al., (1959) 'Cancer and coronary artery disease among Seventh Day Adventists', *Cancer,* 12, 1016–25.

Xiaocong, Wu (1994) *The Subterranean Army of Emperor Qui-shi-huang,* China Travel Tourism Press, Beijing.

Yang, C. K. (1961) *Religion in Chinese Society,* University of California Press, Berkeley.

Yule, H. (1913–15) *Cathay and the Way Thither, Being a Collection of Medieval Notices of China,* vol. 2, Hakluyt Society Publications, London.

Zaehner, R. C. (1970) *Concordant Discord: the Interdependence of Faiths,* Clarendon Press, Oxford.

Ziegler, P. (1970) *The Black Death,* Penguin, Harmondsworth.

Zuk, G. H. et al.(1961) 'Maternal acceptance of retarded children: a questionnaire study of attitudes and religious background', *Child Development,* 32, 525–40.

Index

abbeys, 63, 130
Abdul Aziz Ibn Saud, 155
Aborigines, Australian, 133, 182
abortions, 45, 150
Abraham(ic), 39, 75
Abruzzi, W. S., 135
Ackerknecht, E. H., 169
Adam, 123
Adams, R. Mc., 18
Addis Ababa, 107
Addison, S. T., 58
Afghan(istan), 62, 120, 132, 155
Africa(n), 5, 22, 66–7, 75, 80, 82, 84–6, 88, 91, 99–100, 110, 119, 121, 140, 144, 152, 155, 167, 169, 174, 176, 180, 182–3, 187, 199, 201
afterlife, 18, 58–61, 63, 186
Aga Khan, 154
Agra, 69
Agrawal, B. C., 180
Agrawal, S. K., 180
Ahmaddiya, 154
AIDS, 14, 152, 181, 189
Aladura Independent Church, 152
Albania(n), 46
Albert, E. M., 135
Alexandria, 165
Algeria, 44
Alhambra, 62
Al-Kindi, 169
Alley, K. D., 186
All Souls' Day, 60–1, 89, 187
Alps, 73
Alsace, 144
altar(s), 7, 89
Altekar, A. S., 165
Alther, L., 190
Amara, 142
America (includes North, Central and South), 7, 17, 80, 84–6, 93, 99, 119–20, 124, 131, 135, 143–4, 157, 163, 179, 189–91
Amerindian, *see* Indian (American)
Amiens, 80
Amish, 4, 9, 29, 40–1, 46, 83, 100, 128, 144, 146, 161–2, 164, 181, 207
Ammar, H., 142
Amritsar, 38, 67
amulets, 5, 13, 22, 168
Amundsen, D. W., 169
Anabaptists, 144
Anderson, J. N. D., 141
Andes, 18, 125, 128
Andrews, E. D., 131
Angkor, 19, 33, 64, 119
Anglican Church, *see* Church of England
animals, 4, 7, 22, 30–1, 33–4, 36, 54–5, 63, 75, 91, 98, 105–6, 113, 116, 133, 142, 161, 182–3, 193–203, 210
see also specific names
Ansari, N., 164
Anuradhapura, 20, 193
Arabs, Arabia, Arabic, 5–6, 12, 40, 42, 44, 50, 55–6, 66–7, 75, 81–2, 85, 106, 115, 119, 125, 140, 149, 153–5, 158, 167
Arafat, plain of, 116
Arbellot, G., 175
Arberry, A. J., 62
Ardha Kumbh, 121
Argentina, 146–7
Arizona, 76
Arlington National Cemetery, 124
Armenia(n), 118, 155, 192
Arrington, L. J., 134
art(ifacts), religious, 173
Asoka, 4, 32
Asshur, 123
Assisi, 81, 109
Asso, I. de, 157
Aswan, 92, 97
Atharvaveda, 31
Atlas Mountains, 99
Au Agama, 78

Augustinians, 143
Australia(n), 106, 133, 182
Auvergnats, 74
Avebury, 97
Ayodha, 31, 47, 205
Ayurvedic medical system, 31
Ayuthya, 100
Azores, 11
Aztec, 19, 110

Baal Hammon, 166
Babel, Tower of, 19
Babylon, Babylonia, 11, 97, 123
Baghdad, 4, 93
Ba'hai, 12
Bahakim, H. M., 151
Bahrain, 106
Bali, Balinese, 24, 50, 78–9, 83, 96, 141, 210
Balkans, 85
Banaras, 5, 13, 30–2, 56, 92, 98–9, 116–17, 121–2, 128–9, 138, 186, 197, 205
Banerjea, A. C., 99, 121
Banerjee, A., 139
Bangkok, 35, 79, 95, 109, 173, 193
Bangladesh, 142
Bantu, 203
baptism, 2, 9, 115
Baptists, 138
Bardet, J-P., 176
Barth, F., 36
Bartlett, H. H., 55
Basel, 191
Basham, A. L., 36
basilicas, 117
bats, 193–4
Baxi, U., 139
Bayly, S., 128, 139
Beardsley, R. K., 141
Bedikian, A. Y., 151
Bedu, Beduin, 5, 50, 67, 116, 125
behavioural research, 2
Behrend, H., 174
Beijing, 25, 58, 93, 97, 102, 113, 185
Belfast, 139, 158
Belgium, Belgian, 41, 107
Bellah, R. N., 131
Beloch, K. J., 175
Benedictines, 4, 10, 107, 130, 137, 145, 207
Bengal, Bengali, 30–1, 107, 198
Benigni, U., 60
Bennett, D. H., 106, 146
Bentley-Taylor, D., 32
Berbers, 44
Beresford, M., 145
Bernard, Saint, 48
Bhadavadgita, 28
Bhardwaj, S. M., 120–1
Bible, the, 39, 41, 45, 66, 83, 92, 158, 163, 167
Bieler, L., 176
Bikanir, 195
Binsbergen, W. M. J., 110
birds, ix, 34, 76, 98, 109, 191, 193–5
birth control, 181
birth rate, 104, 150, 172–5, 181, 189
bison, 55
Black Death, 103, 152, 175
Black Madonna, 91
Boardman, E. P., 153
bodhisattvas, 25
Bogoras, W., 201
Boissonnade, P., 129, 143
Bolivia, 127
Bologna, 119
Bolton, J. M., 200
Bombay, 178, 183
Bosnia, Bosnian, 47
Bow (London), 90
Bownas, C., 27
Boxer rebellion, 27, 49
Boyajian, D. H., 155
Boymans-van Beuningen Museum, 185
Brabo, F. X., 147
Bradfield, R. M., 77
Bradford, J., 91
Brahmans, 29, 30, 55, 74, 136, 180, 197
Brandon, S. G. F., 200
Brass, P. R., 156
Braudel, F., 157
Brazil, 146–8, 166, 173
Brethren, 80
Brethren, Exclusive, 2, 12
Brise-Barre, A. M., 8

Britain, British, British Isles, 3, 14, 59, 80, 83, 85, 93, 97, 105, 107, 118–19, 130, 137–8, 155, 158, 167, 196, 206
Brittany, Breton, 46, 97
Brook, S., 140
Brooklyn, 53
Bucking, B., 27
Buddha, 5, 17, 21, 32, 53, 55, 79, 100, 123, 125
Buddhism, Zen, 107
Buddhist, Buddhism, ix, 1–3, 6–7, 9–11, 15, 17–18, 20, 24–5, 28–9, 31–4, 45–6, 49, 58–9, 63–4, 73–4, 78–9, 85, 93–5, 99, 101–5, 107–9, 111–12, 119, 123–5, 133, 137–8, 145, 172–3, 175, 180, 184, 187, 189, 193, 195, 200, 206–7, 210
buffaloes, 54–5
Burma(ese), *see* Myanmar
Bury St Edmunds Abbey, 109, 130
Bush, L. E., 152
bushmen, 166
Bushnak, A. A., 98
Byland Abbey, 130
Byrd, R. C., 163
Byzantium, Byzantine, 42, 82

Cady, J., 49
CAFOD, *see* Catholic Aid
Cairo, 11, 21, 62, 119, 184, 205
Calcutta, 98, 198
California, 153, 162
Calvinism, 74
Cambodia, 4, 33, 64, 79, 210
Cambridge, 80
Canaan, Canaanites, 74
Canada, Canadian, 10, 29, 41, 55, 86
cancer, 161–2, 164
Canterbury, 61, 109, 115, 145
Capellmann, C. F. N., 179
Caraman, P., 147
Carmarthen, 119
Carr-Saunders, A. M., 166
Carthage, Carthaginian, 166
Carthusians, 130
caste (Hindu), 28–9, 32, 38, 48, 128–9, 134, 136, 138–9, 167, 178, 189, 206, 209
Castile, *see* Spain
cathedrals, 5, 14, 63–5, 80, 89, 92–3, 107, 109, 115, 138
Catholic, *see under* Roman Catholic
Catholic Aid (CAFOD), 170
Caucasus, 62, 73
Cavendish, J. C., 45
cemeteries, 5, 27, 47, 59–60, 88–90, 93–4, 109, 124, 139–40, 159, 184, 186–7, 196, 204–5
centralization (of religious groups), 10–11, 25, 28
ch'I, 58–9
Chagga, 131
Chagnon, N. A., 166
Chandrasekaran, C., 173, 178
Chandrasekhar, S., 178
Chartres, 68
Chaudhuri, Nirad, 180
Chavannes, E., 123
Chavis, M., 45
Cheops Pyramid, 17, 20–1, 98, 185
Chernela, J. M., 148
Chiang-Mai, 6, 101, 103
Chidester, D., 124
China, Chinese, 7, 18, 24–8, 32, 49, 58–9, 63, 65–6, 74–5, 87, 89–90, 92–4, 97–8, 100, 104, 109, 113, 117, 123, 132–3, 141, 144, 153, 179–81, 185–8, 193–4, 200, 206, 210
Chipco Movement, 7, 31
Cholula (Mexico), 20
Choudhury, A. R., 31
Christ, 59, 113, 126, 164–5
see Jesus
Christian Aid, 170
Christian Science, Christian Scientists, 128, 152, 179
Christianity, 40ff., 79ff.
see under specific groups
Christians, Indian, 178, 197
Church of England, 64, 119, 137
Church of Scotland, 169
Churches, *see under specific ones*
circumcision, 13, 161, 182
Cistercians, 4, 9, 88, 92, 102, 129, 137, 145
climate, effects of, 24, 50–1, 76–7, 79–81, 110, 154, 168, 208

Coedes, G., 33
Cole, M., 56
Collipp, P. J., 163
Communism, Communist, 8, 64, 93, 132, 158, 185, 188
Confucianism, 25–7, 126
Confucius, 18, 24–5, 194
Congo, 5, 24, 83, 169, 171
Constantinople, 47
contraception, 3, 8, 41, 45, 150, 188–90
Conybeare, F. C., 172
Cook, S. F., 190
Cooke, M. W., 205
Copts, Coptic, 111, 165
Cortés, 190
Cory, H., 178–9
Coulton, C. G., 129–30
cows, 5, 30, 38, 65, 78, 147, 195–200, 202, 206
Coyne, W. D., 119
Cragg, K., 167
Crapanzano, V., 122
Crawford, M. E., 163
cremation, crematorium, 56, 93, 95, 117, 128–9, 185–6
cremation container, 96
Crete, 56, 185, 192
Cribb, R., 158
Croagh Patrick, 123
Croatia, Croatian, 48, 60–1, 187
Cross, H. E., 161
Cross-Upcott, A. R. W., 128
Cruden, A., 167
Crusades, 3, 47, 88, 118, 121, 156, 158
Cyprus, 47, 125, 139, 156, 192
Cyrus, King, 64
Czestochowa, 119

dagobas, 20
Damascus, 98, 120
Damietta, 153
Dandekar, K., 173, 198
Danforth, L. M., 183, 185–6
Daniel, C., 104
Danube, 72
Darlington, C. D., 164, 167
Darwin, Darwinian, 170
David, King, 115
de Witte, Emmanuel, *see* Witte, Emmanuel de
dead, disposal of, 182ff., 205
death rate, 150, 164, 178
deer, 193, 195
deforestation, 13
Deglopper, D. R., 25
Delhi, 14, 108, 115
Derbyshire, 104, 111
deserts, 74–5
dharma, 28, 32, 136
Diana, Princess, 114
Dickemann, M., 167
diet, dietary restrictions/laws, 3, 45, 81, 91, 106, 162, 173, 211
Dinkas, 52, 199
disabled, 167ff.
diseases, infectious, 14, 161, 164, 192
Djarkarta, 98
dogs, 34–5, 104, 109, 198, 200
Dome of the Rock, 10
Dominicans, 144
Dondog, C. R., 172
Donkin, R. A., 145
Dopsch, A., 130
Dorff, E. N., 181
Dorset, 97
Douglas, M., 39
Druzes, 155
Dubai, 106
Dublin, 92
Dubois, J. A., 167
Dubos, R., 81, 129
Dugdale-Caley, W., 131
Dukes, E. J., 59
Dumont, L., 167
Dupaquier, J., 104, 176
Dupont-Sommers, A., 49
Dutch, *see* Netherlands
Durham, 80, 109
dwarfism, 164

East Anglia, 93
East Indies, 72
Eck, D. L., 117, 121
ecumenical movement, 7, 156
Eddy, Mary Baker, 127, 152
Efrat, E., 149

Egypt, Egyptian, 4, 10, 14, 17–19, 22, 44, 50, 61, 67, 75–6, 94, 100, 117–18, 133, 142, 154–5, 160, 184–5
Eickelman, D. F., 120
Eickhorn, W., 27
Eid el-Kabir, 119
Eid es-Saghir, 119
Eire, *see* Ireland
Ekvall, R. B., 124, 200
Eleazar ben Jair, 191
El Tayeb, E. M., 121
Elgood, C., 169
Ellen, R., 141
Ellison, C., 163
Elvin, M., 63
Encyclopedia Britannica, 28, 176
Encyclopedia Judaica, 155, 161, 175, 192
England (ish) 4, 18, 60, 63, 66–7, 73, 90, 93, 97, 107, 125, 127, 129–30, 132, 137, 140, 145, 150, 171, 175, 177, 182, 184–5, 191
Enlightenment, 66
Enstrom, J. E., 162
environmental management, 204–8
Erech, 123
erosion, 13, 91, 208
eskimos, 50, 53, 81, 84, 87, 166
Ethiopia, Ethiopian, 3, 11, 89, 105–6, 108, 112, 118, 151, 165
Europe, European, 3–4, 7–8, 12, 22, 40–1, 44, 47, 49, 52, 54–5, 63, 66–7, 77, 80, 84–5, 89, 94, 107, 114, 119–21, 124, 128–9, 137, 143, 148, 154, 156, 170, 175
Eusebius, 165
Evans-Pritchard, E. E., 52, 65, 87, 177
Everest, Mount, 123
Evers, J. C., 54
exclusivity (religious), 12–13
excreta, disposal of, 57
Eyam, 104

Falashas, 151
Falun Gong, 132
Farooq, M., 164
fast, fasting, 105–6, 166
Fatima, 99, 122
feng-shui, 18, 25, 58–9, 186
Fenland, 145
Ferngreen, G. B., 169
fertility, human, 9
Field, M. J., 168
fish, 34, 76, 79, 81, 105, 143, 193
Five Pillars of Islam, 88, 207
Flanders, 187
Flick, A. C., 191
Florence, 119
Foggia, 91
folk religion, 28
foods, use of, 104ff.
Forbidden City, *see* Beijing
forests, forestry, 90, 110, 128, 132, 134, 143, 185, 194
Fortier, B. de la, 170
Fountains Abbey, 109, 130
Fox, George, 127
Francaviglia, R. V., 135
France, French, 55, 60, 73, 85, 104, 113, 118, 144, 171, 176–7
Franciscan Missionary Sisters, 169
Frank, J. A., 179
Frankfurt-am-Main, 191
Fraser, G. E., 162
Freed, R. S., 198
Freed, S. A., 198
frogs, 34, 76
Fromageau, G., 104, 176
Fulop-Miller, R., 147
fundamentalists, 11, 140, 205
funeral, funerary, 2, 19, 33, 60, 89, 92, 94, 185–6
Furnivall, J. S., 129

Gadgil, M., 90
Gajdusek, D. C., 106
Galanter, M., 165
Gamer, H. M., 105
Gandhi, Mahatma, 7, 29, 152, 196, 206
Gandhi, Mrs, 38, 48
Ganesh, 30
Ganges, Gangetic, 30, 32, 56–7, 78, 90, 99, 117, 119, 122, 129, 186, 198
Ganshof, F. L., 130
Garner, J., 163
Garonne, 145
Gaul, *see* France
Gautama, 32, 49

Geller, S. A., 184
Geneva, 74
Genoa, 119
Gentiles, 161
geology, 80
geomancers, 58
German, German, Germanic, 56, 85, 130, 182, 192
Ghana, Ghanaian, 127
ghats, cremation, 30
ghettos, ghettoization, 40, 46, 85, 139, 148, 158, 192
Giles, L., 174
Glasse, R., 106
Goa, 189
Gorer, G., 188
Gothic, 80
Gourou, P., 199
Grader, C. J., 78
Graham, A. C., 27
Granada, 62, 156, 157
Grant, C. P., 98
graves, *see* tombs
graveyards, *see* cemeteries
Gray, R. F., 76
Greece, Greek, 47, 60, 63, 74, 89, 123–4, 157, 185–6, 192, 194
Greenland, 84
Greenoak, F., 194
Griscom, J. H., 162–3
Groen, J. J., 107
Groussert, R., 153
Guildford, 64
Gujarat, 7, 38, 48, 139, 199
Gutkind, E. A., 89
Guyana, 165
see Jonestown

Hadith, 127
Haigh, C., 175
Hajj, ix, 5, 43, 83, 88, 91, 98, 113, 115–16, 120–2, 125, 151, 200–1, 204, 207
Hall, J. R., 165
Hamarneh, S., 169
Hamas, 158
Hands, H. R., 63
Haney, C. A., 170
Hangchow, 193
Hanoi, 58, 113
Hanuman, 30
Hare Krishna, 7
Harijans, *see* Outcastes
Harrell, D. E., 163
Harris, M., 32, 197
Hartley, I. S., 163
Hartley, Robert, 162
haruspiciation, 201
Hashemites, 140
Hawaii, 86
Hayes, J., 105
health, 160ff
Heber, R., 117
Hebrew, 56
Heer, F., 191
Herodotus, 118
Herskovitz, M. J., 199
Hickey, J., 158
Himalayas, 53, 73–4, 125, 175
Hindu Kush, 74
Hindus, Hinduism, 1–3, 5, 7, 10, 12–15, 17–18, 28–35, 46–9, 56–7, 75, 77–8, 82–3, 85, 90, 93–4, 99–101, 104–5, 114–17, 119, 121, 123–4, 126, 128–9, 134–6, 138–9, 150–1, 156, 158, 165, 167, 169, 173, 178, 180–1, 184, 187, 189, 193–8, 200, 205–7, 209–11
Ho Chi Minh, 113, 187
Hodgetts, T. J., 205
Holiness Church, 106
Holmberg, A. R., 91
Holocaust, 3, 192
Holy Ghost Fathers, 169, 171
Hong Kong, 59, 105, 186
Hopi, 66–7, 76–7
Hospitallers, 121
Hostetler, J. A., 144, 182
Huguenots, 191
Hunt, Holman, 68
Huntington, G. E., 182
Hurd, J. P., 181
Hutterites, 4, 10, 29, 40–1, 75, 146, 162, 164, 175, 181, 207
hygiene, 56, 81, 100, 160–1, 163, 176, 182, 208

Ibadan, 98

Ibn Saud, 155
Iceland, 3
ihram, 116
Iknaton, 76
imams, 61
immunization, 139, 179, 181
inbreeding, 46, 181
Incas, 97
India, Indian, ix, 4, 5, 7, 14, 20, 24, 28–32, 35, 38, 47, 55, 57, 74, 82, 88, 90, 94, 99, 104, 106, 114, 118, 120–1, 128–9, 132, 136, 138–9, 151, 154, 156, 161, 165–7, 173, 180–1, 189, 193, 195–9, 205
Indian (American), 54–5, 76, 97, 123, 144, 146–8, 157, 173
Indonesia, 14, 79, 96
Indo-China, 18
Indus, 18
industrialization, 8, 195, 211
Inoguchi, R., 165
Iona, 73, 119
Iran, Iranian, 5, 10, 12, 38, 42, 63, 74, 82, 118, 132, 153–4, 156, 191
Iraq, Iraqi, 12, 61, 63, 65, 94, 142, 154, 156, 191–2, 207
Ireland, Irish, 41, 46, 56, 73, 85, 99, 119, 123, 158
Irrawaddy, 20, 113
irrigation, 24, 50, 57, 78, 81, 144, 156
Isaac, E., 21
Isbell, B. J., 128
Isfahan, 153
Islam, Islamic, ix, 1–12, 17–18, 28–9, 31, 39–44, 61–3, 66, 80–2, 85, 87–9, 94, 104, 115, 122, 126–7, 132–3, 142–3, 151–2, 154–5, 156, 160, 167, 169, 173–4, 177–8, 180, 184, 187, 204–5, 207–8, 211
see also Hajj, Mecca, Mohammed, Moslem
islands, effects of, 86, 95, 99
Ismailis, 154
Israel, Israeli, 4, 10, 12, 40, 62, 85, 99–100, 106, 124, 140, 148, 151, 154, 158, 175, 192, 196, 200–1, 204
Istanbul, 61
Italy, 60, 73, 81, 91, 113, 119, 175, 177
Ivory Coast, 4, 65, 93

Jacobi, H., 37
Jains, Jainism, 9, 17–18, 28, 31, 34, 36–7, 49, 53, 78, 105, 123, 145, 200, 210
Jaipur, 53
Japan, Japanese, 4, 7, 18, 24, 27–8, 32, 59, 73, 89, 95, 100, 113, 124, 129, 141, 165, 178, 188, 193, 198, 210
Java, 20, 32, 158
Jedda, *see* Jiddah
Jehovah's Witnesses, 2, 128, 153
Jerome, 164
Jerusalem, 7, 13, 92, 98–9, 113–15, 122–4, 140
Jervaulx Abbey, 130
Jesuits, 10, 15, 73, 86, 130, 146–7
Jesus, 55, 208
see Christ
Jews, Hasidic, 2
Jews, Jewish, 3, 6–9, 11–13, 39–40, 43, 45, 48–51, 53, 61, 74, 85, 99–101, 104, 106–7, 109, 113–14, 123–4, 126, 128, 134, 140–1, 148–9, 155, 158, 160–1, 164–5, 169, 175, 181, 184, 189, 191–2, 208
Jiddah, 98
jihad, *see* war
Jonestown, 3, 124
Jordan, 4, 132, 154
Joseph, M., 171
Judaism, Judaic, 1, 17–18, 39–41, 55, 74, 115, 126, 164–5, 170–1, 173, 179–80, 184, 207–8, 211
Juergensmayer, M., 158
Jumna, 99

Ka'aba, 67, 89, 110, 116, 122
Kadhimain, 61
Kaffirs, 74
Kakar, S., 138
Kaku, K., 178
Kalahari, 5
Kamba (Kenyan tribe), 177
Kamikazi pilots, 165
Kansu, 89
karma, 28, 48, 136
Karnak, 19, 64
Kashmir, 122, 138
Kazemzadeh, H. K. Z., 120

Kenya, 52, 169, 171, 184
Kenyon, S. M., 174
Kerala, 136
Kerbela, 61
Keuneman, H., 193
Keys, A., 107
Khambageu, 76
Khartoum, 155
Khurajaho, 173
kibbutz, kibbutzniks, 9, 85, 148–9
kiKamba, 67
Kilimanjaro, Mount, 131
Kimambo, I. N., 90
Kindere, L. V., 187
Kirkstall Abbey, 130
kismet, 141
kiSukuma, 56
kiswahili, 66
Kitchener, General Herbert, 155
Knesset, 13
Knight, H., 198
Knights of St John of Jerusalem, *see* Hospitallers
Knights Templars, *see* Templars
Knock, 61, 99, 119
Knutsson, K. E., 105, 165
Koran, 39, 42, 55, 62, 66, 81, 92, 104, 127, 154, 158, 160, 167, 172, 208
Korea, 32, 74, 95
Kosovo, 12, 46
Krakow, 192
Krantz, G. S., 55
Krech, S., 54–5
Krishna, 30–1, 197
Kumbh Mela, ix, 5, 99, 119, 121
Kuwait, 207
kwoth, 201
Kyoto, 19

Lachiver, M., 104, 176
Laidlaw, J., 37
Lakshmi, 197
lamas, 124, 172
Lamb, T., 153
Lambo, T. A., 168
landscape, 2, 41, 44, 54, 58, 68, 78, 81, 94
Lane, B. C., 124, 134
languages, 56, 65–7, 92, 197, 204
Lansing, J. S., 78
Lao Tse, 18, 24
Larson, D., 162
Latter-Day Saint(s), 2, 4, 10, 81, 91, 118, 124, 134–5, 152–3, 161–2, 182
lazar houses, 170
Leach, E., 21
Lebanon, 12, 47, 155, 158, 174
leprosy lepers, leprosarium, 161, 170–1
Levin, J. S., 162–4
Levine, R. R., 179
Levy, G., 43, 48–9
Lhasa, 64, 79, 113, 125
Liberia, 140
Libya, 44
Lidenthal, E. L., 124
Lindisfarne, 73
Liverpool, 4, 64, 93, 138
Livingstone, D., 77
Lodrick, D. O., 197, 199
Loftin, J. D., 77
Loizos, P., 56
London, 5, 8, 46, 59–60, 90, 93, 132, 187
longevity, *see* death rate
Loos, 130
Losito, P., 164
Lourdes, 61, 88, 97–9, 110, 112, 121–2, 125, 150–1, 163, 205
Ludwig, E., 52
Luig, U., 174
Lutkins, S. G., 188
Luxor, 18–19, 64, 92, 97, 100

Macinnis, D. E., 58, 186
Mackenbach, J. P., 107
Maddison, D., 188
Madeira, 100
Madras, 93, 121
Maghreb, 43, 61
Mahabharata, 28, 121
Mahayana, 32
Mahdiyya (Sudan), 155
Malawi, 22
Malaysia, 200
Maliki School of Law, 177
Mallam, M. B., 164
Malta, 121

Manchester, 98
Mandalay, 115
Mandelbaum, D. G., 199
Manila, 15, 27, 59, 114, 187
Manley, G. T., 196
Manu, Laws of, 180
Mao Tse-Tung, 113, 187
Maputo, 166
marabouts, 61
Marnham, P., 121
Marrakech, 4, 93
marriage, x, 2, 9, 12, 177, 180–1, 191, 206
  levirate, 173
Marsh, G. P., 2
Marshall, P., 63, 177
Mary, Virgin, 91, 95, 112–14, 158
Masada, 165, 190–1
Masai, 52, 91, 123, 184, 199
Maspero, H., 27
Massacre of St Bartholomew's Night, 191
materials, use of, 91ff.
Mayadas, C., 198
Mayans, 20
Mbuti, 24
McEvoy, L., 162
McGovern, T. H., 84
McGuire, R. B., 163
McKusick, V. A., 46, 164
McMillen, S. I., 160
McNaspy, C. J., 147
McNeil, J. T., 105
materialism, 18
Mecca, 3, 8, 10–11, 42, 47, 51, 55, 67, 83, 89, 91, 98, 110, 115–22, 125, 150–1, 174, 201–2, 205
Medina, 42, 55
Medjugorje, 99, 112, 122
Megasthenes, 194
Meller, H., 90
memorials, 89, 93–5, 124, 188
Memphis, 18
Mencius, 24
Mendel, Gregor, 145
Mennonites, 9, 40, 209
Mercier, C. A., 170
Merida (Mexico), 146
Merton, R. K., 153
Meslin, M., 27
Mesoamerica, Meso-America, 17–19, 97, 117
Mesopotamia, 17–18, 22, 123, 154, 192
Methodism, Methodists, 5
Mexico City, 98, 121, 144
Mexico, Mexicans, 19–20, 86, 91, 97, 109, 143–4, 146, 190, 200
Midgely, M., 195
Milan, 119
Milner, M., 28
Mina, 116–17
Ming Dynasty, 185
ministers, 81, 87
Mishkat-al-Masabih, 62
Mishnah, 181
Mississippi, 179
Mitford, N., 186
Moguls, 47
Mohammed, 6, 42, 55, 62, 82, 91, 113, 115, 122, 125, 154, 167, 174, 180, 208
Mohenjodaro, 18
Mombasa, 9, 177
monasteries, monasticism, 7, 9–11, 25, 34, 40, 49, 64, 81, 88, 92–3, 102, 107, 109, 113, 119, 124, 127, 129–31, 137, 143, 145–6, 150, 172, 193, 206
Mongolia, 32, 79, 172
monks, 32, 40, 49, 63–4, 73, 79, 87–8, 92, 102–3, 107, 109, 129–31, 143, 145, 172, 175, 180, 200, 210
monkeys, 30, 34–5, 193–5, 210
monotheism, monotheists, monotheistic, 74–5, 93
Montezuma, 190
Montreal, 170
Moors, 156
Morenz, S., 76
Moriscos, 157
  see Morocco
Mormons, *see* Latter-Day Saints
Morocco, Moroccan, 120, 122, 154, 157
Mosaic injunctions, 160, 200
moshav, 148–9
moshav shitufi, 148–9
moshava, 148

Moslems, 2–3, 5, 7–8, 10–12, 14, 32, 42–4, 47–8, 50–1, 56–7, 61–2, 65–6, 82–3, 85, 87, 91, 94, 100–1, 104–5, 114, 116–17, 119, 123–5, 128, 132, 136, 138–42, 151–3, 155–6, 161, 164, 166, 174–6, 178, 184, 192, 195–7, 200–1, 206–7
mosques, 4, 5, 8, 11, 14, 30, 45, 47, 50, 58, 61, 63, 83, 86–9, 93, 108–9, 115–17, 158
Moulay Idris, 122
Moule, A. C., 185
Mount Olympus, 123
mountains, effects of, 25, 50, 73, 78, 80, 90, 95, 98, 111, 122–5, 131–2
Mubarak, A., 142
Mukerjee, R., 198
Murra, J. V., 201
Mus, P., 33
Museum Boymans-van Beuningen, *see* Boymans-van Beuningen Museum
Muzdalifah, 116
Myanmar, 3, 15, 20–2, 32, 34, 50, 52, 63, 91–3, 100, 104, 108, 111, 113, 129, 146, 173, 200, 206, 208
Mysore, 136, 167

Nakajima, T., 165
Nanak Shah, 37
Nandra, P. N., 197
Nansen, F., 166
Nantes, 191
Nara, 193
Narmadi Dam Project, 7
Nash, J., 128
natman, 59
Nayars, 166
Nazis, 40, 182, 192
Neech, A. S., 196
Needham, J., 194
Nejd, 82
Nejev, 61
Nelson, H. G. H., 186
Nepal, 32, 124, 175
Netherlands, 41, 107, 185, 191
New England, 134
New Guinea, 106, 165, 177, 201
New Mexico, 24, 76, 135, 144, 206
New Testament, 158, 167, 208
New York, 2, 4, 40, 64, 93, 162–4
Newsome, A. E., 106
Nicaragua, 95
Nigeria, 85, 121
Nile, 52, 75–6, 108, 184–5
Nilgiri Hills, 199
Nilotic tribes, 199
nirvana, 54
Nodari, R., 177
Nolan, M. L., 120
Nolan, S., 120
Noonan, J. T., 176
North Carolina, 190
Northamptonshire, 194
Noss, J. B., 27
Novotny, T., 179
Nuer, 24, 52, 65, 87, 177, 199, 201
nuns, 3, 40, 74, 83, 87, 172, 175, 187
Nuremberg, 191

observatories, 87
O'Hara, M. L., 169
Ohsawa, G., 107
Old Testament, ix, 39–40, 50, 52, 158, 167, 208
Oliver, S., 154
Olivier, L., 164
O'Malley, P., 158
Olmecs, 97
Omari, C. K., 90
Orang Asli, 200
Orni, E., 149
Orsi, R., 64
Orthodox Christian, 192
Ortiz, A., 76
Orubuloye, I. O., 176
ossuaries, 89
Ottoman Empire, 152
Our Lady of Guadaloupe, 121
Outcastes, 29, 48, 57, 75, 85, 124, 136, 178, 197–8
Oxford, 80

Pagein, 20
pagodas, 16, 20, 33–4, 93–4, 107–8, 111, 113
Pakistan, 4, 14, 31, 104, 106, 132, 151, 154, 204

Pakrasi, K. B., 167
Palestine, Palestinian, 6–7, 12, 47, 99, 115, 121, 124, 140, 149, 158, 191, 211
Palmer, M., 25–6
Pandey, G., 138
Pannikar, R., 30
Panter-Brick, C., 152
papal, *see* Pope
Paraguay, 9, 146–7
Pare, 90
Paris, 5, 46, 113
Parish, W. L., 188
Parrack, D. W., 57, 197–8
Parry, J. P., 47, 56, 99, 129
Parsis, 9, 38–9, 178, 183
pastoralism, pastoralists, 44, 50, 79, 81, 149
pastors, 88
Patimokkha, The, 49
PDSA, 196
Pearl Harbor, 124
Peel, J. D. Y., 152
Peires, J. B., 203
Peking, *see* Beijing
Pembrokeshire, 97
Persia, *see* Iran
perceptions of the environment, 53–71
persecutions, 9, 12
Peru, 46, 86, 91, 97, 110, 184
Philippines, 14–15, 23, 50, 208
Phillips, R. L., 162
Philp, H. R. A., 169
pigs, 52, 104, 161, 195, 201
Piggott, J., 27
pilgrims, pilgrimages, 13, 30, 33, 43, 61, 84, 86, 92, 95–9, 103, 109, 115–27, 130, 151, 163–4, 174, 195, 201–2, 204–5
Pillars (of Islam), 88, 207
Pirovane, G., 177
Pitcairn Island, 81
planning, 15–16
Plymouth Brethren, 128
Pogrom, 155
Poland, 4, 155
Pollitzer, R., 121
pollution, polluted, ix, 56–7, 78, 91, 107, 109, 117, 121, 136, 186, 208
polyandry, 175
polygyny, polygynous, 104, 118, 134, 150, 174, 176, 181–2
Pope, the, papal, 127, 152
populations, 1, 4, 8, 12, 15, 22, 51, 74, 81, 84–5, 116–18, 135, 145ff., 153–5, 164, 167, 170, 172, 175–6, 180–2, 186–7, 189, 192, 201
pork, 45, 161, 195
Portugal, Portuguese, 29, 148
Postan, M. M., 130
Potala Palace, 64, 79, 113, 125
Potter, R. G., 104
Pousin, L. de la V., 59
Prescelly Mountains, 97
Presley, Elvis, 187
priests, priesthood, 3, 9, 43, 74, 83, 87–8, 128, 130, 147, 157–8, 169, 172, 175, 177, 188, 198
Pritchard, J., 76
Protestants, ix, 7–9, 41, 46, 49, 56, 85, 99, 109, 137, 139, 151, 185, 209
Prussia, 191
Puzo, M., 176
Pyrenees, 113
pyramids, 17, 19, 21, 93, 97–8, 110, 184–5

qibla, 11
Quakers, 128, 162
Quaraish, 42
qubbas, 61
Quint, M. N., 142

rabbis, 9, 88, 99
Rackham, O., 90
Rafeq, A. K., 120
rain(fall), *see* climate
Rajasthan, 34
Ramadan, 3, 43, 87–8, 91, 105, 115, 119, 166, 173
Ranger, T. O., 169
Rangoon, 48
Rao, P. S., 181
Rapoport, D. C., 158
Rappaport, R. A., 201
Rasmussen, K., 166
Ray, D. J., 87

Reading, 61, 109
Reddy, D. V. S., 169
Rees, W. D., 188
Reformation, 63, 110
Regent's Park Mosque, 8
Rewell, R. E., 161
Reynolds, V., ix, 1, 195
Rhodes, 121
Rhodes, Cecil, 130
Rhodesia, *see* Zimbabwe
Ricard, R., 144
rice, 23, 52, 78, 113, 129, 208
Richards, I. A., 66
Richardson, J. F., 155
Rievaulx Abbey, 109, 130
Robbins, J., 54
Roberts, D. F., 9, 107
Robinson, R. H., 32
Rogers, L., 57
Roman Catholics, Roman Catholicism, 2–3, 7–8, 15, 29, 41, 45, 49, 56, 61, 63–5, 73–4, 83, 85, 88–91, 93, 99–100, 104, 108–11, 113, 117, 119, 121, 123, 127–8, 137, 139, 143–6, 148, 151, 157, 164, 169–70, 176, 178–9, 181, 187–9, 191, 207, 209
Rome, Roman, 6, 63, 80, 98, 109, 118, 156, 164, 184, 191
Romsey Abbey, 130
Romsey Marsh, 145
Rosen, G., 163
Rosnay, L. de, 130
Ross, J. L., 175
Rostock, 191
Rotterdam, 185
Rouen, 51
Roy, P., 189, 198
Roy, T. K., 189, 198
RSPB, 196
RSPCA, 196
Rudner, D. W., 129
Rupesingh, K., 155
Russell, C., 154
Russell, W. M. S., 154
Russia, Russian, 3, 12, 132, 155, 172, 190, 192

Sabbath, sabbatical, 128, 134
Sachar, H. M., 140
sadhus, 173
Sahara, Saharan, 52, 106, 191
Sahel, 52
St Albans, 61, 109, 115
St Anastasia, 127
St Andrew, 127
St Benedict, 129
St Bernard, 48
St Francis (of Assisi), 81
St Germain des Près, 130
St Giles, 127
St Joseph, 127
St Stephen, 127
St Thomas, 123
Salesian Mission, 148
Sallnow, M., 141
Salt Lake City, 118, 124
Salvation Army, 171
Samarra, 94
Sangren, P. S., 96, 122
Sanscrit, 77, 206
Santiago de Compostela, 61, 99, 119
Sapir, E., 66
Saragossa, 61
Saraswati, B., 117
Sattin, A., 30
Saudi Arabia, 4–5, 8, 10, 87, 93, 98, 109, 116, 132, 151, 204
Savoy, 74
Scandinavia, 84
Schafer, E. H., 194
Schapera, I., 178
schizophrenics, 168
Schwarz, B. E., 106
Schwedagon, 190
Schweitzer, Albert, 170
Scotland, Scot, 3, 22, 73, 137
Scribner, S., 56
Scurletis, T. D., 170
Seattle, Chief, 54
segregation, female, 14, 57, 178, 192
Seguy, J., 144
Seine river, 191
Selinus, R., 105, 165
Semple, Ellen C., 53, 82, 86, 118
Sepoys, 106
Serbia, Serbs, 46

Seventh-Day Adventists, 75, 81, 83, 106, 128, 153, 162
sexual rules, sexual intercourse, 172ff.
Shafi School of Law, 177
Shakers, 100, 127, 131, 134, 172
Shang-Ti, 25
Shanghai, 63, 205
Shantung, 185
Sharot, S., 99
Sharp, L., 133
Shastri, C. P., 197
sheep, 104, 130, 144, 147, 201
Sheffield, C., 9
Sherring, M. A., 166
Shi'ites, 3, 11, 61, 63, 142, 151, 154–5, 180, 191
Shilluks, 52, 199
Shintoism, Shintoists, 17, 27, 73
Shiva, 48, 123
shrines, 5–7, 10, 13, 19, 21, 23, 57, 78, 89–90, 100, 102–3, 108, 110–15, 121, 124, 127, 140, 151, 163, 195
Shwedagon Pagoda, 107
Siberia, 201
Sicily, 125
Siena, 175
Sierra Leone, 140
Sikhs, Sikhism, 28, 37–8, 48, 67, 83
Sikkim, 188
Silbury Hill, 97–8, 125
Silesia, 129
Sima, 98
Simoons, F. J., 105, 200
Sims, G. R., 90
Sioux, 203
Siva, 30
  see also Shiva
Skar, S. L., 141
Skopsi, 3, 172, 190
Slack, P., 152
Smith, H. S., 76
Smith, W. R., 95, 167
Social Science and Immunization Research Council, 139
Soga, T., 202
soil fertility, 144
Solomon, N., 40
Somalia, 182
Somerset Levels, 145
South Africa, 2, 202–3
Southwold, M., 63
space, use of, 88ff.
Spain, Spanish, 85–6, 129, 143, 147, 156–7
Spencer, Stanley, 68
Spier, F., 46
spirits, 10, 22–5, 56–8, 95, 110–13, 123, 127, 131, 133–4, 141, 174, 188, 194, 201, 208
Spiro, M. E., 16, 33, 52, 63, 79, 187
Sprigg, J., 134
springs, 90, 97, 110, 112, 141, 206
Sri Lanka, Sri Lankan, 4, 11–12, 20, 30, 32, 46–7, 63, 74, 85, 92, 99, 112, 123, 125, 138, 193, 195, 210
Srinivas, M. N., 128, 136, 197
Ssonjo, 57, 76–7
statues, statuary, 33, 80, 95, 100, 113–14, 127–8
Steiner, G., 67
Stevenson, H. N. C., 91
stone circles, 97
Stone of Scone, 22
Stonehenge, 18, 97
Stover, L. E., 141
Strikman, M., 27
stupas, 20, 36, 94, 193
Suakin, 98
sub-castes, Hindu, 46, 128–9, 167, 197, 209–10
Sudan, 24, 47, 50, 52, 61, 85, 87–8, 121, 155, 174, 182, 199, 201
Sukuma, 2, 7, 23, 52, 56, 113, 131, 179, 200–1
Sumatra, 32
Sunni, 3, 142, 154–5, 191
Suras, *see under* Koran
Surbey, M. K., 175
suttee, 165
Swahili, 92
  see kiswahili
Swaroop, S., 121
Sweden, 85
swidden agriculture, 24, 90
Swiss, Switzerland, 9
synagogue, 86, 102, 109, 140, 163
Syria, 75, 154, 192

Venezuela, 148
Venice, 80, 119
Verhulst, A., 130
Vermont Agricultural Society, 2
Vietnam, 32, 79
Viola, A., 188
violence, violent, 9, 49, 151, 153ff., 156, 158, 190–2, 202, 205
Virgin Mary, *see* Mary, Virgin
Virginia, 179
Vishnu, 30, 48
Vogt, E., 135
Voland, E., 179
Volney, *see* Rafeq

Waco, 3
Waghorne, J. P., 93
Wahhabis, 82, 109, 155
Wailing Wall, 11, 113–14
Wakeman, F., 153
Waldenses, 73
Wales, 119, 145
Walland Marsh, 145
Wallerstein, I., 93
Walsingham, 119
waqfs, 10, 63, 169
wars, warfare, 154, 158, 170, 175–6, 178–9, 191
Warsaw, 192
Washington, 54
Waugh, E., 186
wealth, effects of, 63ff.
weather, *see* climate
Weber, M., ix
Wei Ming, 66
Weiner, I., 161
Welch, H., 26, 64
Welsh, *see* Wales
Werbner, R. P., 127
Werner, D., 173
Wesley, John, 110
West, D. W., 162
West Kennet Long Barrow, 125
Westminster Abbey, 113
Weyl, N., 166
Wheatley, P., 19, 58
White Fathers, 169
White, Ellen Gould, 127
Whorf, B. L., 66
Whyte, M. K., 188
Wilkinson, D., 27, 49, 155
Willcocks, Sir William, 132
Witte, Emmanuel de, 185
Wilson, G. E., 153
Wolf, E. R., 15
World Council of Churches, 7
World War First, 177
World War Second, 8, 12, 155, 165, 192
World Wildlife Fund, 195
Wright, Q., 155
Wuthnow, R. A., 163
Wynder, E. L., 161–2

Xhosas, 202–3
Xian, 19, 87, 93–4, 98, 109, 185

Yang, C. K., 49
Yanomamo, 166
yin and yang, 26, 27
York, 165, 190
Yorkshire, 130
Young, Brigham, 127, 134
Yucatan, 146
Yule, H., 194

Zaire, *see* Congo
zakat tax, 43
zar cults, 174
Ziegler, P., 152
ziggurats, 11, 19, 89, 97, 123
Zimbabwe, 16, 130
Zoroaster, Zoroastrianism, 28, 38, 74
Zuk, G. H., 179
Zuni Indians, 10, 131, 144

Taipei, 27
Taiping Rebellion, 27, 153
Taiwan, Taiwanese, 89, 95, 122, 186
Taj Mahal, 69
Takeda, C., 188
Taliban, 132, 155
Tamurlane, 153
Tanganyika, *see* Tanzania
Tanner, A., 54
Tanner, R. E. S., 1–2, 9, 23, 46, 50, 52, 58, 104, 113, 195
Tanzania, 7, 22–4, 49, 52, 56, 76, 83, 88, 90, 113, 123, 125, 128, 131, 143, 178, 184, 200–1
Tao, Taoism, Taoist, 25–7, 100, 109, 122, 126, 133
Tao Te Ching, 26
Tapia, A., 190
TEARFUND, 170
Templars, 48, 121
Temple Mount (Jerusalem), 123
temples, 4–5, 11, 14, 16–17, 19, 21, 25, 27, 30, 33, 35–7, 52, 63, 78–9, 83, 86, 88, 91–3, 102, 109, 115, 124, 136, 141, 145, 173, 188, 194–5, 204, 210
Teotihuacan, 19
terrorism, 158
Tewa, 76
Thailand, Thai, ix, 3, 5–6, 32, 35, 79, 88, 92, 100–1, 103, 113, 133, 173, 206
Thebes, 18
Theravada, 32
Thirty Years War, 3, 7
Thomas, E. S., 59
Thompson, S. E., 152
Tibet, 32–3, 64, 74, 79, 113, 124–5, 175, 200, 206
time, use of, 86ff.
tirtha, 78
Todas, 199
Toelken, B., 77
Tokyo, 27, 59, 94–5, 113, 125
Toland, J., 165
tombs, tombstones, 4, 89, 97, 109, 113, 124, 140, 183–5, 194, 204
Topley, M., 105
Torah, 160–1
Tower Hamlets (London), 90
transhumance, 42
transmigration (of souls), 28, 31, 49, 59, 206, 209
transport, use of, 96ff.
Trappists, 107
Troyer, H., 162, 164
Tuan, Y-F., 123
Tuareg, 5, 191
Tunisia, 44, 154
Turkestan, 154
Turkey, Turks, Turkish, 47, 56, 85, 152, 155, 192
Turnbull, C., 24, 110, 123, 152
Turner, E., 120, 123
Turner, H. W., 152
Turner, V., 115, 120, 123
turtles, 117

U Nu, 145
Uganda, 110, 123, 152, 169, 181, 190
Ukaegbu, A. O., 174
Ulster, 12, 41, 139, 157
Underwood, P., 164
Underwood, Z., 164
United Arab Emirates, 120
United Nations, 205
United States of America, 10, 29, 45, 52, 86, 88, 95, 106, 157–8, 161–2, 169, 179–80
  see also America
Universities Mission to Central Africa, 169
Untouchables, 129, 138–9
Upanishads, 28
Ur, 19, 94, 97, 123
USA, *see* United States of America
Utah, 10, 134, 153, 161–2
Uttar Pradesh, 99

Valley of the Kings, 19
Vanderpool, N. Y., 162
Vartak, V. D., 90
Vatican, 88, 105, 117
Vedas, 28
Veer, Pn der, 139
vegetarian, vegetarianism, veganism, 9, 31, 36–7, 78, 81, 101, 105, 107
veiling (of women), 10